OXFORD TEXTBOOK OF FUNCTIONAL ANATOMY

VOLUME 1

REVISED EDITION

by PAMELA C. B. MacKINNON
and JOHN F. MORRIS

Department of Human Anatomy, University of Oxford

OXFORD UNIVERSITY PRESS
1994

Oxford University Press, Walton Street, Oxford OX2 6DP

Oxford New York Toronto
Delhi Bombay Calcutta Madras Karachi
Kuala Lumpur Singapore Hong Kong Tokyo
Nairobi Dar es Salaam Cape Town
Melbourne Auckland Madrid
and associated companies in
Berlin Ibadan

Oxford is a trade mark of Oxford University Press

Published in the United States
by Oxford University Press Inc., New York

A catalogue record for this book is available from the British Library

Library of Congress Cataloging in Publication Data
MacKinnon, Pamela C. B.
Oxford textbook of functional anatomy/by Pamela C. B. MacKinnon
and John B. Morris.—2nd ed.
   p.   cm.—(Oxford medical publications)
Includes index.
Contents: v. 1. Musculoskeletal system.
1. Human anatomy. I. Morris, John F. II. Title. III. Series.
[DNLM: 1. Anatomy. QS 4 M158]
QM23.2.M33   1993   611—dc20   92-48317
ISBN 0 19 262195 5

Typeset by Cotswold Typesetting Ltd, Gloucester
Printed in Hong Kong

OXFORD TEXTBOOK OF FUNCTIONAL ANATOMY

VOLUME 1

Foreword

By Charles G. Phillips BSc, MA, DM, FRCP, FRS
*Formerly Dr Lee's Professor of Anatomy in the University of Oxford;
and Emeritus Fellow of Trinity and Hertford Colleges, Oxford*

In my final years as a University teacher–researcher I had the great pleasure of collaborating with Pamela MacKinnon and John Morris in supervising the practical work of first-year undergraduates who were learning living anatomy under the guidance and within a framework of notes written by these two experienced teachers of anatomy—notes which they were improving every year in response to the encouragement of the undergraduates, and which were gradually taking shape as a book that could be found interesting and useful by a wider public.

The General Medical Council's Recommendations have been liberating in two ways: first, in encouraging Universities to diversify their approaches to medical education instead of compelling them, as hitherto, to conform to a common pattern; and secondly, in designating detailed topographical anatomy as a postgraduate study. A detailed knowledge of topographical anatomy, unless refreshed by daily practical application, can be held in memory only by exceptional individuals; and since a proportion only of medical undergraduates are intending to, or will ever, become surgeons, it is important that medical undergraduates in general should not have their education distorted by spending excessive time and energy on dissection of the whole body. The old idea that every newly qualified doctor should be ready to perform major surgery on any kitchen table has been dead for many years. Aspiring surgeons must nowadays spend two or more postregistration years before being entrusted with independent responsibility, and during those years they must make sure of all the topographical detail they have to know. As for preclinical education, more than lip service has now to be paid to the idea that biomedical science is advancing so rapidly that continuous reorientation will be needed throughout professional life, and that an essential minimal core of unchanging fact will have to be carefully defined.

Thus the question arises: what should be the content and emphasis of first-year anatomy? Different schools will—and should—make different approaches to the problem. The course that is being evolved by Drs MacKinnon and Morris and their colleagues is firmly centred on familiarity with the structure and macroscopic functions of the living body, investigated as far as possible with eye and hand complemented by modern imaging techniques, clinical and electrical work on the action of muscles, study of prosections made by surgeons-in-training, and careful dissection of limited regions of the body. Above all, the objectives of first-year anatomy are clearly proclaimed in their book. Without definite guidance it is all too easy for undergraduates to suppose that they are expected to memorize the whole contents of the great atlases and encyclopaedias of anatomy. This course is, rather, a training in accurate observation and systematic recording: the observing to continue, and the records to accumulate, throughout professional life. Every first-year anatomist must understand and remember the principles of first aid. Beyond this, only exceptional people can remember all they need to know without ever having to refer to their books and personal records. Such records ought obviously to be consulted by every clinical student who knows he will have to present a case on a ward-round. And every doctor who takes part in an expedition to outlandish places ought obviously to prepare her- or himself to intervene, if necessary, in one of the few life-threatening emergencies that might arise.

Acknowledgements

Any textbook of Anatomy must necessarily reflect, in part, the instruction of our teachers and the enthusiasms of our colleagues over the years. To all of these, and in particular Professors the late Sir Wilfred LeGros Clark, FRS, the late G.W. Harris, FRS, Barry Cross, FRS, Charles Phillips, FRS, Ruth Bowden, and Joseph Yoffey, our warm and grateful thanks for their expertise and encouragement.

Some of our colleagues, to whom we are indebted, have made important contributions to particular sections of this book: Mr Eric Denman has advised on many surgical points; Professor Charles Phillips FRS initiated the section on the experimental testing of elemental movements; Dr George Gordon initiated the section on the experimental testing of cutaneous sensation; Drs Basil Shepstone and Stephen Golding provided the chapter on Medical Imaging and many of the radiographs; while Drs Philip Anslow, Paul Chamberlain, Walter Fletcher, David Linsell, Andrew Molyneux, Nial Moore, Daniel Nolan, and Simon Ostlere provided other illustrations of medical imaging. The vast majority of the artwork has been drawn by Audrey Besterman, who valiantly put up with our comings and goings; many of the photographic illustrations were made by the photographic unit at the Department of Human Anatomy, Oxford, under the direction of Brian Archer; Robert Underwood designed and made the electronic apparatus referred to in Chapter 3. Jane Ballinger was responsible for much of the early typing. Many of our colleagues, in particular Professor Chris Hall-Craggs, Dr Margaret Matthews, and Lady Margaret Florey have read and made helpful comments on earlier versions of this book. And last but not least, since all anatomists are dependent on dissected material, we gratefully acknowledge the careful conservation of prosections by Roger White and Terry Richards.

Contents

CHAPTER 1

Introduction

Expectations and use of the book

In writing this book we have deviated from traditional approaches and attempted to emphasize functional and living anatomy and its imaging in medical practice. The book has been designed to enable students of morphology, whether they be medical or dental, physiotherapists or radiographers, nurses, or any other student of the biology of Man, to understand the basic functional design of their own body and that of others.

To dissect an entire cadaver from head to toe can be very instructive and is essential for aspiring surgeons. However, the recent and continuing escalation of knowledge in all branches of the preclinical and clinical medical sciences, for which time must be provided in medical teaching programmes, makes this increasingly impractical in most courses. Nevertheless, the understanding of how the body is built and how it works is fundamental to the practice of all aspects of Medicine.

This book (Vols. 1, 2, 3) therefore guides its readers through the fundamentals of anatomy by use of appropriate prosections and a partial dissection of the body. Each seminar has been designed to cover an aspect of the musculoskeletal system which can be studied in about a 2 h period, in a combination we have found suitable. Different courses use dissection and prosections in different proportions. Dissection by students can well start sooner than is suggested here; the book is not intended as a comprehensive dissecting manual.

Looking at the Contents it might appear that undue emphasis has been placed on the innervation of the upper limb. However, we anticipate that the upper limb will be studied first. Armed with an understanding of the arrangements in the upper limb, a study of the lower limb is considerably easier since the general principles are similar. This is reflected in the smaller number of seminars devoted to the lower limb and in particular to its nerve supply. Although we suggest a study of the spine after that of the limbs, it is equally appropriate if that order is reversed.

For those who do not have access to prosections and the various types of medical images, the book can also be used by referring to the reader's own body and to the illustrations. Liberal reference should be made to embryology, histology, and neuroanatomy texts and the book should be read in conjunction with physiology and biochemistry texts. It is important that you are able to answer the questions which have been inserted throughout the book; if the foregoing text has been absorbed and understood then they should not prove difficult. Personal notes on your anatomical investigations made in the margins of the book will prove invaluable. At some future time, rapid perusal of a seminar containing additional comments of your own will ensure an equally rapid recall of the basic information. This is a book which, if used properly, should never be discarded.

Functional Anatomy is the structure and function of the body in its living state: it is a study of our skeletal system which protects the vital organs and gives attachment to muscles; of muscles and joints which provide for movement between the various skeletal units; of the highly specialized cardiovascular system through which oxygen and nutrients are pumped to individual cells of the body and waste materials are collected for excretion; of the various organs within the head and neck, thorax, and abdomen which enable the body to remain viable by ensuring a homeostatic environment for each individual cell; of the reproductive system which ensures continuity of the species; and of the endocrine and nervous systems which receive and integrate information from both the internal and external environments and which, by the secretion of hormones or neurotransmitters, influence the behaviour of the cells of the body to control the functions of all the other systems. In this way the systems of our body maintain that body, and allow it to respond to and interact with our changing environment and, through speech, movements, and behaviour, to express our individual character and personality.

The changing form of the body and its relations to function

Always remember that each body is unique and not an assembly-line product. It develops in the uterus and this development usually produces a 'normal' individual. It develops and grows further during childhood and adolescence in sexually dimorphic growth spurts to produce the adult forms. On occasions, any part of this development may be imperfect to a greater or lesser degree. It will also be obvious to you that variation among normal individuals exists. It is important that you develop a concept of the **range of normality** so that you can judge what is abnormal. For this purpose many illustrations of abnormalities are included in the book. In later adult life, ageing changes lead to senescence. Most bodies donated for examination in the dissecting rooms are those of elderly people.

External differences between male and female are mostly obvious. However, the body of both sexes undergoes many cyclic changes of a circhoral and diurnal nature and that of the female undergoes in addition a monthly cycle. Throughout life, the body responds morphologically to functional demands (e.g. muscle hypertrophy) and to abuses and injuries by repair and healing.

The 'body' which you must consider is therefore not the static, usually elderly form which you see on the dissecting room table, but rather a living dynamic organism, constantly changing and responding to the functional challenges of its environment.

Any region of the body consists of a number of different tissues. The anatomical form that these tissues take in any region is the result of their evolution to fulfil a functional role. The structure of tissues can be considered on two levels. The microscopic structure is the object of study in histology. The macroscopic (naked eye) and radiological structure is the subject of these seminars.

Terms used in anatomical description (1.1, 1.2)

For ease of communication and convenience of description the body is always considered as standing erect, facing ahead, the arms by the sides and the palms of the hands facing forwards with the fingers extended. Place yourself in this position and note that it differs in a number of ways from the way in which you normally stand.

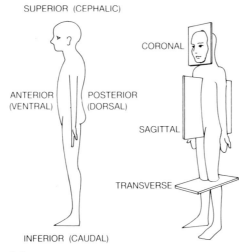

1.1
Anatomical terms and planes.

The terms **anterior** (ventral) and **posterior** (dorsal) refer to structures facing the front and back of the body respectively. Structures in the antero-posterior midline are said to be **median**, and those close to the midline **medial**, and those further away, **lateral**. Structures above are usually referred to as **superior** or **cephalic**, or if they are above and in front as **rostral**. Structures below as **inferior** or **caudal**. **Proximal** means nearer to the origin of a structure, **distal** is the opposite. **Superficial** means nearer the skin, **deep** is the opposite.

Anatomical planes (1.1)

Sagittal—a vertical plane lying in the antero-posterior plane (longitudinal);
Coronal—a vertical plane at right angles to the sagittal; **Transverse**—a horizontal plane at right angles to both coronal and sagittal;
Oblique—any plane that is not coronal, sagittal or transverse.

Movements (1.2)

Movement of a limb away from the midline of the body is termed **abduction**; **adduction** is the movement which returns the limb to its original position or towards the midline of the body. A forward or anterior movement of the trunk or a limb is term **flexion**; a backward or posterior movement is **extension**. A combination of movements of flexion, extension, adduction, and abduction (without rotation) is termed **circumduction**. The forearm can be **pronated** so that the palm faces posteriorly, or **supinated** so that the palm faces anteriorly. If the sole of the foot is directed medially it is **inverted**; if directed outwardly it is **everted**.

Rotation can occur at certain joints. If the rotation is towards the midline of the body the joint or body part is said to be **medially (internally) rotated**; if away from the midline, **laterally (externally) rotated**. These movements at the shoulder can be conveniently demonstrated, respectively, by swinging the arm, with the elbow bent, across the chest and then away from the chest. Rotation of the head on the neck is referred to as **pivoting**.

A forward movement of the head, jaw, or shoulders is called **protraction**; a backward movement is **retraction**.

If the non-rotated back is kept straight and the head or trunk is bent to either side the movement is referred to as **lateral flexion**.

For movements of the thumb and other digits see p. 62.

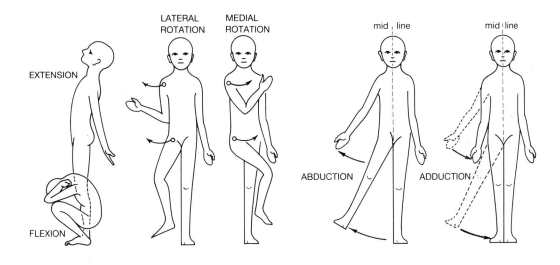

1.2
Movements of the body.

CHAPTER 2

Systematic thinking about tissues

Aims To outline the tissues that make up the musculoskeletal system and limbs and to indicate topics to which particular attention should be paid. This short account must be expanded by a study of the histology of the tissues.

Skin

The skin consists of an epithelium (**epidermis**) on a base of connective tissue (**dermis**).

Consider:

● **skin colour**: black, brown, or yellow skin depends primarily on the amount of melanin pigment secreted by the melanocyte cells of the epidermis. Melanin helps protect the deeper layers of the epidermis from the harmful effects of ultraviolet light.

● **degree of keratinization**: keratinization is protective (compare the sole of the foot and the eyelid) and is affected by the environment, for instance the calluses caused by manual work.

● **nails** are horny plates of modified epidermis which cover the dorsum of the distal phalanges and which provide a firm base for the pulp of the fingers or toes. The growing root of a nail extends as far as the white (lunula) of the nail and is overlapped by a fold of skin (eponychium). The rest of the attached nail appears pink because of the underlying capillaries which supply the nail bed. Nails develop at the tips of digits but migrate on to the dorsum taking their nerve supply with them. Nails of the hand grow faster than those of the foot.

● **degree of hairiness** and **type of hair**: the palms, soles, eyelids, and penis are hairless. The **density** and **coarseness** of the hair differs from region to region. The distribution of body hair is sexually dimorphic, the male pattern being dependent on circulating androgens. Abnormal hair distribution can therefore reflect endocrine imbalance.

● **dermal ridges** of the hands and feet: these function in gripping and in texture recognition (movement of ridged skin over an object produces vibrations that are sensed by cutaneous nerve endings). The pattern of ridges forms the 'finger-print' and 'palm-print' which is unique to an individual (except for identical twins).

● **sweat glands** and their openings: **eccrine** glands are found over most of the body surface in Man and secrete a colourless watery saline. The openings of their ducts can be located on the surface of the dermal ridges. The number of open active sweat glands on most parts of the body can be increased by exercise and are important in the control of body temperature. Increased emotion or attention however causes activity of sweat glands on the palms and soles (see Chapter 3). **Apocrine** sweat glands in the axilla, anogenital region, and mammary areola develop at puberty under the influence of sex hormones. They open into hair follicles and produce sweat with a higher organic content which bacterial enzymes break down, causing odour formation.

● **sebaceous glands** also open into hair follicles. Their cells desquamate to form an oily secretion (sebum) which lubricates the hair and skin. Sebum secretion is enhanced by androgens at puberty and is associated with acne. The **areolar glands** of the nipple are specialized sebaceous glands which become prominent in pregnancy and lactation.

● **skin creases**: at these pronounced lines, which are found especially around joints, the skin is more firmly attached to the underlying tissues. Distinguish skin (flexure) creases from the fine crease-like lines which appear in the skin of old people and which are caused by degeneration of collagen fibres and reduced attachment of the skin to underlying tissues.

● **skin cleavage and tension lines**: The bundles of skin collagen fibres, the underlying muscles, and the fibrous bands between the muscles and the overlying skin all have a particular orientation. Therefore, if a round puncture wound is made in the skin the resultant skin defect is often elliptical. On the face the tension lines are marked by skin wrinkles. In general, the lines of maximal tension run circumferentially round the trunk and limbs (**2.1**). Incisions made *parallel* to lines of maximal tension heal with a fine, inconspicuous scar; incisions *perpendicular* to tension lines produce a more hypertrophic scar.

● **blood supply of the skin**: this is derived from local vessels. Capillary loops in the nail bed can be seen if the skin is cleared with a drop of oil and observed with a dissecting microscope. The thermal sensitivity of skin vasculature can readily be demonstrated by immersion in hot and in iced water. In cold conditions much of the cutaneous blood supply is short-circuited from arterioles to venules by arterio-venous anastomoses.

● **innervation of the skin**: stimuli to the skin differ in terms of their energy (e.g. mechanical or thermal), their spatial distribution, intensity, and rate of change. Many cutaneous sense organs are highly sensitive to mechanical and others to thermal stimuli, and yet others are less specific. Some respond rapidly and transiently, others in a more sustained manner.

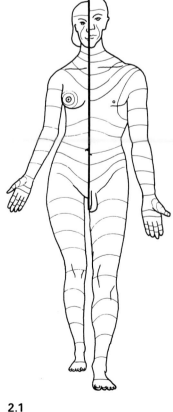

2.1
Skin tension lines.

Differences in packing density of the receptors produce large differences in the spatial discriminatory power of different regions. This is usually measured by the two-point discrimination threshold—the smallest distance between two simultaneously applied mechanical stimuli that can be perceived as two rather than one (see Chapter 3).

You will need to know:

- the **local nerve** which supplies the area of skin: for diagnosis of nerve lesions and for administration of local anaesthesia.

- the **spinal nerve** which supplies the area of skin: for diagnosis of spinal cord and spinal nerve lesions. The area of skin supplied by a spinal nerve is called a **sensory dermatome** (see p. 29).

Superficial fascia

Superficial fascia is the subcutaneous connective tissue which merges with the dermis of the skin. It consists of an aqueous matrix in which are fat cells, fibroblasts and their bundles of collagen fibres, plasma cells, mast cells, and macrophages. These vary considerably in amount from area to area. It therefore forms a compartment which either tethers the skin or allows it to move over the deep tissues, and provides a store of energy (fat), and cells which protect against invasive organisms.

Consider:

- the **fibrous tissue** content: this determines the attachment of the skin to deeper structures. Compare the palm (where the skin is tightly bound to facilitate grip) with the dorsum of the hand. Fibrous attachments are prominent at skin creases, and form the suspensory ligaments of the axilla, breast, and penis. In areas such as the anterior abdominal wall, distinct fibroelastic sheets are present.

- the extent of **fat** deposition and its regional variation: compare the thigh and abdomen with the eyelids, dorsum of hand, and penis. Most subcutaneous fat is 'white' adipose tissue in which adipocytes store fat in single large droplets. Its distribution becomes sexually dimorphic at puberty; its extent is largely dependent on the balance between food intake and energy expenditure. Some 'brown' adipose tissue is found in newborn humans. Its cells contain many small fat droplets and mitochondria. It is well supplied with capillaries and sympathetic nerves and provides a rapidly available source of energy.

- the presence of **fluid-filled** sacs: these subcutaneous **bursae** allow the skin to move freely over bony prominences.

- the **superficial vessels and nerves** which pass through the fascia to reach the skin. A superficial system of veins, lymphatics, and nerves runs within the superficial fascia, with fine terminal branches of arteries.

Deep fascia

Deep fascia is a dense fibrous connective tissue that covers and ensheathes muscles and is attached to bones. In some parts the skin is strongly attached to it. It provides sites for muscle attachment, forms partitions (intermuscular septa) separating muscle groups, and bands (retinacula) which hold tendons in place. In most regions it is mainly fibrous, but always contains some fat and fluid. Its thickness varies considerably. Over expansile organs such as the pharynx it is very thin, but in areas such as the leg it forms a non-expansile sleeve which is important in the mechanics of venous return. Many larger texts name the deep fascia according to the muscle that it covers, but only a few of these are important to remember.

Bone

Bone is a vascular connective tissue which owes its great strength to a matrix containing collagen fibres on which crystals of calcium hydroxyapatite have been laid down. In most mature bone the collagen fibres are laid down by osteoblasts in osteons (Haversian systems) which consist of concentric lamellae of matrix and its cells (osteocytes) surrounding a central canal containing vessels and nerves and osteoclasts (cells which break down bone). In the earliest bone and in bone formed immediately after fractures, the collagen fibres are randomly arranged, forming 'woven' bone. **Compact** bone forms the cortex, and most of the shaft of long bones. Within most bones is a marrow cavity which is braced by **trabeculae**, bony struts arranged along lines of stress. Bone is continuously remodelled and can thus adapt to changing environmental stresses.

Development of bones. Bones are formed from the mesoderm of the embryo. They all develop first as a mesenchymal (loose mesoderm) model. In most cases, the mesenchyme chondrifies to form a cartilaginous model of the bone, which later ossifies (**ossification in cartilage**). By contrast, the clavicle and bones of the skull (except its base) ossify in models composed of condensed mesenchyme (**ossification in membrane**).

Bone formation starts at a **centre of ossification** which frequently lies centrally in the soft tissue model and then spreads centrifugally. Many bones, particularly smaller bones (such as those of the carpus, tarsus, and auditory ossicles) develop from a single centre. These centres appear over a wide period ranging from the 6th week of intrauterine life to the 10th postnatal year, appearing later in the smaller bones.

Many other bones (in particular the long bones and girdle bones of the limbs) are ossified from multiple centres. The **primary centre** usually appears near the middle of the bone (in long bones, the middle of the shaft) early in development, from the 6th (clavicle)–16th week *in utero*. From this primary centre ossification proceeds towards the ends or periphery of the bone. **Secondary centres** appear at the ends (periphery) of the bone from a time just before birth (the centre for the lower end of the femur) onwards to late teenage. This is the mode of ossification of the long bones, metacarpals, metatarsals and phalanges, the ribs, and the clavicle. Between the ossifying shaft (**diaphysis**) and ends of the bone (**epiphyses**), plates of cartilage (**epiphyseal plates**) remain and, at these sites, growth continues. Linear growth is usually more marked at one end of a long bone (the '**growing end**') than at the other. Growth gradually ceases towards the end of puberty and the epiphyseal plates become ossified so that the shaft fuses with the epiphysis. This normally occurs earlier in girls than in boys. Fusion of the epiphyses to the shafts is usually complete by 18–21 years. Increases in the girth of a bone are caused by **appositional growth** in which bone is deposited by osteoblasts beneath the periosteum. At the same time, the medullary cavity is increased in size by erosion of the endosteal surface of the bone by osteoclasts.

During the entire growth period, the anlage and the developing bone are being continually remodelled to 'keep pace with' the growth. Varying tensions on bone, such as are exerted by the insertions of muscles, tendons, or ligaments, alter the contour in that area, e.g. the deltoid tuberosity. Abnormal tension or pressure can lead to gross deformity.

Knowledge of the times of appearance, rates of growth, and times of fusion of the secondary centres is often needed in clinical practice, for instance in assessing the skeletal age of a child in comparison with its chronological age, or in forensic medicine, but these can be looked up at the appropriate time. A radiographer will often take an image of both the normal and the abnormal limb on one film to assess the symmetry.

Individual bones

You should be able to identify any bone and hold it in the **position** that it occupies in the living body.

Consider:

- the position of its **articular surfaces** and the bones with which they articulate.
- **named parts** and **prominences**, especially those that are palpable in the living subject.
- the site of major **muscle attachments** which may or may not be associated with roughened areas or protrusions of the bone.
- the site of major **ligament** and **membrane attachments**.
- the **blood supply** and position of nutrient arteries.
- the **marrow cavity** content of red or fatty marrow; its extent in children and adults.
- any specializations of the **trabecular** pattern within the bone or thickenings of the cortex which reinforce particular lines of stress in the bone.
- the **ossification** of the bone—and whether ossification occurs in a **membrane** or **cartilage** model.

Cartilage

Cartilage is an avascular connective tissue which forms the articular surfaces of synovial joints, the cartilaginous models of developing bones, and the pliable skeleton of the nose, pinna, and larynx. Its cells (chondrocytes) lay down a resilient, very hydrated matrix which is rich in collagen, and aggregates of glycosaminoglycans such as hyaluronic acid with proteoglycans (**hyaline** cartilage). Other forms are particularly rich in **fibrous** or **elastic** tissue.

Joints and their movements

Joints are the articulations between bones. They therefore consist of different types of tissue. Their form varies widely in relation to the functional requirements of the articulation. The degree of **mobility** varies widely: some joints permit virtually no movement; at others, small gliding or angular movements occur; and at many, a large range of movement in different planes can occur. Consider the factors that will influence (a) the **type of movement** that can occur, and (b) the **range of movement**. In any joint, the movement that occurs will depend on the deformability of the material uniting the bones and on the force that muscles can exert to produce that deformation. In addition, all joints must be **stable**. In joints which have considerable mobility, always consider the specializations which give stability to the mobile structure.

Classification of joints

Joints can be classified in a number of ways. The most common depends on the material that separates the bones (fibrous tissue, cartilage, or a synovial cavity). Joints may also be classified according to the type or extent of movement that can occur and whether the joint is temporary or permanent.

A. Fibrous joints

The bones are united by fibrous tissue. The extent of possible movement depends on the length of the fibrous tissue in relation to its cross-sectional area. If it is long, as in the sutures of the skull of a child at birth or the interosseous membrane between radius and ulna, then considerable movement is possible. If it is short, as in the sutures of the adult skull, the peg and socket joints between the teeth and jaws, or the interosseous ligament of the inferior tibiofibular joint, then little movement can occur.

B. Cartilaginous joints

In '**primary' cartilaginous joints** the bones are united by hyaline cartilage. Such joints occur temporarily between the epiphyses and diaphyses of long bones and permanently in other places such as between the first rib and the manubrium. A little flexing occurs between first rib and manubrium during respiration (Vol. 2) but no movement should occur at epiphysial plates provided that the structure of the cartilage is sound (but see p. 96).

In '**secondary' cartilaginous joints** the articulating bony surfaces are covered with hyaline cartilage and united by fibrocartilage. Such joints all occur in the midline of the body: between the bodies of the pubic bones (pubic symphysis); between the bodies of the vertebrae (intervertebral discs); and between the manubrium and sternum. A little movement occurs at all these joints, but endocrine-induced changes in the pubic symphysis allow greater movement during late pregnancy and parturition.

C. Synovial joints

In synovial joints the bones, covered with articular cartilage, are separated by a fluid-filled synovial cavity. These are the most common joints in the body. Many allow a considerable amount of movement between the bones, but at some virtually no movement occurs. Further classification of synovial joints is based on the shape of the articular surface: **plane**; **hinge**; **pivot**; **condylar**; **ellipsoid**; **saddle**; **ball and socket**. These descriptions are, of course, only approximations; all articular surfaces are ovoid to some degree. In one position the articular surfaces are most congruent (close-packed) and the joint is most stable.

For any joint consider:

• the **nature** and **range** of the **movements** possible. This depends on:
– the shape of the articular surfaces;
– the deformability of the tissues uniting the bones;
– restrictions by ligaments;
– restrictions by soft tissue apposition (e.g. the arm against the trunk in adduction of the shoulder);
– the mechanical advantage of muscles crossing the joint.

All movements can be categorized as either **sliding**, **rolling**, or **spinning** of one joint surface on the other. These fundamental motions are combined to produce the movements at a joint.

Movements at a joint can be classified as:

Flexion—Extension
Lateral flexion
Abduction—Adduction
Medial (internal)—(external) Lateral
rotation of limbs
Rotation (of head, trunk)
Pronation—Supination
(of the forearm)
Inversion—Eversion
(of the foot)
Protraction—Retraction
(of shoulder, head, jaw)
Elevation—Depression
(of the shoulder, jaw)
Opposition of the thumb (and little finger)

• the **stability** of a joint. This depends on:
– the articular surfaces;
– ligaments;
– muscles.
Of these, only the muscles provide *active* support. If they are paralysed, the ligaments will soon stretch and joint deformity will ensue.

The structure of a joint reflects an evolutionary compromise between mobility and stability in relation to the function of the joint.

• the **articulating surfaces**: the bones that take part, and the shape of their articular surfaces. In some cases the areas of contact in different movements is important.

• the **capsule**: its extent, attachments, strengths and deficiencies. The capsule is fibrous and usually attached to the *margins* of the articular surfaces. Note where it deviates from this.

• the **ligaments**:
– **Intrinsic ligaments** are thickenings of the capsule laid down along particular lines of stress.
– **Accessory ligaments** also limit movement of the joint but are separate from the capsule.

• the **synovial cavity** and **synovial membrane**: the synovial membrane usually lines all the non-articulating surfaces within a joint. The amount of fluid depends on the contours of the bones within the cavity; but some larger incongruities are taken up by mobile fat pads covered with synovial membrane so that the actual volume of fluid is very low, and is scarcely more than a molecular layer of fluid between the articulating surfaces. The fluid has thixotropic properties i.e. when its molecules are under pressure cross linkages break down and the fluid becomes less viscous. On radiographs, the 'space' between the congruent articulating bones is occupied largely by the articular cartilage, which is radiolucent.

• any **bursae**: some are extensions of the synovial membrane out of the joint capsule which provide fluid-filled sacs to give friction-free movement of e.g. tendons over bones. Others related to the joint may not connect with the synovial cavity.

• **intra-articular structures**: **fat pads** covered with synovial membrane help to spread synovial fluid. **Intra-articular discs** of fibrocartilage divide the cavity of certain joints in which movements occur in two distinct axes.

• the **blood supply** to a joint: there is usually a good anastomosis of the arteries around joints which have a large range of movement, and many of the local arteries give branches to the capsule. This arrangement provides for a continuous supply distal to the joint even if the position of the joint tends to reduce the flow in some of the larger arteries.

• the **nerve supply** to a joint: the capsule of joints has an important **sensory** nerve supply. This conveys **mechanoceptive** information to the central nervous system concerning the direction, rate, and acceleration of any movement of the joint. It also signals excessive movement via **pain** fibres. A nerve which supplies a muscle acting over a joint also supplies sensory fibres to that joint. More specifically, a nerve supplies that part of the capsule which is made slack by the contraction of the muscles it supplies. For instance, the nerve to biceps also supplies the anterior part of the capsule of the elbow joint, so overstretching of the capsule can be prevented by reflex contraction of biceps. An intact nerve supply is thus essential to protect a joint from traumatization due to excessive movement.

Vasomotor fibres of the sympathetic nervous system supply arterioles of the synovial membrane.

Muscles and their actions

Muscle tissue is formed from elongated cells which contain filaments of the proteins actin and myosin which interact to produce tension when intracellular calcium is increased. In **skeletal** or **striated** muscle the myofilaments are arranged in register in parallel longitudinal bundles; hence the microscopic striation. Skeletal muscle contracts rapidly in response to nervous stimulation; 'white' fibres contract particularly rapidly, 'red' fibres contract more slowly but are less easily fatigued and are thus more common in postural muscles. **Smooth** (non-striated) muscle occurs in the walls of blood vessels and many viscera. Its filaments are not arranged in parallel bundles. Its fibres contract slowly and in a more sustained manner; different subtypes depend more or less on their innervation for stimulation and their cells are arranged, respectively, in less or more of a functional syncytium. **Cardiac** muscle shares many features of striated muscle, but its cells are branched and linked mechanically and electrically so that the heart contracts as a coordinated whole.

Many details of the topography of skeletal muscles are not of clinical importance, though there are obvious exceptions such as the positions of the tendons of the wrist in trauma surgery, and the arrangement of the muscle layers around the inguinal canal in understanding hernias. It is generally much more important to understand the **muscle groups** that produce given movements at a joint and their innervation.

Consider:

● the **action** of a muscle in a movement. This may be classified as:
– **prime mover** (agonist): produces the required movement.
– **synergist**: prevents unwanted movements which would be produced if the prime movers acted alone.
– **essential fixator**: clamps certain parts in a position on which the movement of other parts is based.
– **postural fixator** (e.g. of the trunk): prevents the body being toppled by movements of heavy parts which shift the centre of gravity.
– **antagonist**: opposes the prime movers in a particular movement. During the movement they are normally relaxed in proportion to the power of the prime movement.
– 'paradoxical' actions counter the force of gravity (e.g. biceps contracts when an elbow is extended while lowering a heavy weight).

A particular muscle may be a prime mover in one movement, an antagonist in another, a synergist or essential fixator in others. Depending on the relationship of the attachment of a muscle to the joint over which it acts, the major component of the tension generated by the muscle may act to produce the movement, or to maintain the articular surfaces in contact while the movement occurs (**2.2**). Muscles with a chief role as **prime movers** therefore tend to be attached so that they have a considerable degree of mechanical advantage (e.g. biceps); muscles with a primary postural function tend to be shorter and more closely applied to a joint. The muscles of the spine illustrate this particularly well (p. 148).

● The **attachments** of skeletal muscles are usually to two separate bones so that the muscle crosses a joint. They may have additional attachments to fibrous tissue but the details are usually unimportant. Some muscles (especially in the face) are attached to the skin. The **origin** of a muscle is described as its more proximal attachment, or the attachment that usually remains fixed during the prime movement produced by the muscle. The **insertion** is the more distal attachment and the part that usually moves. The terms are really only used for convenience of description since muscles act differently in different movements.

An individual muscle fibre can contract by no more than one-third of its length. A muscle can therefore originate no closer to its insertion than the point at which the maximum excursion of the joint would require a *ca* 30% shortening of its fibres. Thus muscles around very mobile joints must originate further away. Consider, for example, the origin of the short scapular muscles from the medial two-thirds of the scapular fossae. Fibres attached closer to the joint could not shorten sufficiently to allow the full range of medial and lateral rotation without buckling (**2.3**).

The **method of attachment**: some muscles appear to be attached directly to bone, but a small amount of fibrous tissue (microtendons) always intervenes. Others are attached via **tendons** (rounded bundles of fibrous tissue) or **aponeuroses** (flattened sheets of fibrous tissue). These allow: (i) the bulk of a muscle to be separated from its point of action; (ii) the pull of a muscle to be concentrated into a small area; and (iii) the line of pull of a muscle to be altered.

● the **shape** of a muscle and the **arrangement of its fibres**: two basic principles govern the form of a muscle. Both determine the angle at which the muscle fibres are arranged:
– The maximum degree of shortening is proportional to the length of the muscle fibres.
– The maximum power of a muscle is proportional to the number of muscle fibres.

Muscles with parallel fibres (**2.4a**) can shorten most but, for a given volume of muscle tissue, the number of fibres can be increased by placing them

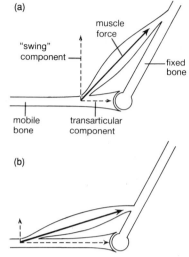

2.2
Components of muscle action at a joint in relation to their site of attachment. (a) muscle inserted close to joint; (b) muscle inserted distant from joint.

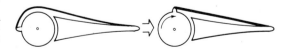
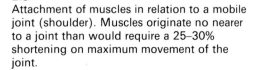

2.3
Attachment of muscles in relation to a mobile joint (shoulder). Muscles originate no nearer to a joint than would require a 25–30% shortening on maximum movement of the joint.

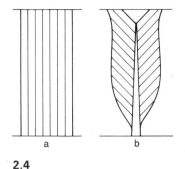

2.4
(a) Parallel and (b) pennate
arrangement of muscle fasciculi.

obliquely to the direction of pull (**2.4b**), although this reduces the length of fibres and thus the degree of shortening possible.

Muscles are often described according to their form. Although the function (and thus the form) of muscles must be understood, memorizing the form of individual muscles is not necessary, though a few examples of different types should be borne in mind.

- the **nerve supply** to muscles. You will need to know the **local nerve** of supply to a muscle to diagnose peripheral nerve injuries. Understanding the **spinal segmental nerve** supply (myotome) is important in the diagnosis of spinal nerve and cord lesions, but is best learned for muscle groups (Tables 1, 2; pp. 76, 131). All muscles producing a given movement of a joint have the same segmental spinal nerve supply (e.g. elbow flexors are supplied from C5, C6) and the antagonists are supplied by adjacent spinal segments (elbow extensors are supplied from C7, C8). **Posterior primary rami** of spinal nerves supply the extensor muscles of the spine; all other muscles are supplied by **anterior primary rami**. Flexor and extensor muscles (the two basic embryological groups) in any limb are supplied, respectively, by anterior and posterior **divisions** of the anterior primary rami of spinal nerves in the plexuses supplying the limbs.

The **size of the motor units** (number of muscle fibres supplied by one nerve axon) determines the precision of action possible. For example, muscles producing finger movement have small motor units while those in the buttock muscle gluteus maximus are very large.

- the **blood supply** to muscles: Muscles need a good blood supply, but the details of the local arteries of supply are rarely of importance. In general, adjacent arteries supply a muscle, and there is a major point of entry of the neurovascular bundle. The two ends of a muscle receive a local supply.

The vasculature

The vessels in which blood and lymph circulate consist of three layers: an endothelium (*tunica intima*) which provides a friction-minimizing lining to the vessel and, with its basement membrane, the exchange surface of capillaries; vessels other than capillaries also have a *tunica media* which consists of elastic, fibrous, and smooth muscle tissue in varying proportions; and an outer layer (*tunica adventitia*) of fibrous tissue in which run the vessels and nerves which supply the larger vessels.

Arteries

The arteries form a high-pressure distribution system. The largest arteries have a very elastic tunica media to absorb the pulsatile force of the heart contraction; smaller arteries have a more muscular coat. The **arterioles** which form from them and which regulate blood flow to a part, have a smooth muscle coat that is large in relation to the diameter of the lumen and is controlled by sympathetic nerves.

For any artery, consider:

- its **origin** and parent vessel.

- its **course**, particularly where its pulsations are palpable, or where the vessel is exposed to injury. Arterial pulses are palpable only at certain points; usually where they cross bone. Such sites can form useful 'pressure points' where digital pressure can arrest distal haemorrhage. The course of large vessels can readily be mapped in the living by using an ultrasonic Doppler-based probe. Radiologically, vessels can be demonstrated by the injection of radio-opaque material into the circulation at an appropriate point (angiography) (p. 19).

- its **major branches** and **mode of termination**.

- its **area of supply**.

- the **extent of anastomosis** with other major vessels. Some vessels, such as the central artery of the retina, are **end arteries**; some are *functionally* end arteries because of very limited anastomoses; yet others have plentiful anastomoses.

Vessels of the retina can be observed directly in the fundus of the eye by use of an ophthalmoscope; capillary loops can be visualized in the nail bed. The capillary bed, arteriovenous anastomoses, and other small vessels of the circulatory system cannot be visualized with the naked eye.

- the **degree of oxygenation** and **haemoglobin content** of the blood can best be assessed where the skin is thin, at such places as skin creases in the palm and the conjunctiva of the eyelids.

Veins

The veins form a low-pressure capacitance system. In the limbs a system of **superficial veins** is separated from the **deep veins** by the deep fascia. Like arteries, veins are lined by endothelium. In smaller veins below the heart the endothelium forms **valves** which break up the hydrostatic column of blood. The tunica media is thinner than that of arteries because pressure in the veins is much less.

For any vein consider:

- its **mode of commencement**, remembering that this is distal.

- its **course**, particularly where the vein can be punctured with a needle for intravenous administration of substances or withdrawal of blood for testing; also areas which are liable to trauma.

- any **valves** within the veins, noting the marked regional variation in the incidence of valves.

- the **mode of termination** and **major tributaries** of the vein.

- the **area drained** by the vein.

- the **extent of anastomosis** with other veins, paticularly the **communicating** veins which link the **deep** and **superficial** veins of the limbs through the deep fascia.

Lymphatics

The lymphatic system is a very-low-pressure system that returns extracellular fluid, proteins, and cells to the blood vascular system. It resembles the venous system except that the endothelium of its blind-ending capillaries is discontinuous and therefore its vessels are more permeable. They are also much smaller and have more valves. The walls of lymphatic capillaries are connected to the surrounding connective tissue in such a way that the presence of excess extracellular fluid pulls on the connections and opens the vessels. The blood vascular system is protected from invasion by microorganisms via the lymphatics by the presence of **lymph nodes** along their course. These are collections of cells of the immune system, which filter the lymph and respond to foreign proteins (antigens) with an immune response. Multiple afferent vessels enter a node; a few efferents leave via its hilum.

Lymphatic vessels cannot be detected by routine examination of a living subject unless they are inflamed or enlarged. Also, they may be difficult to dissect. They can be demonstrated radiologically after injection of radio-opaque dyes. It is, however, crucially important that you are familiar with the lymphatic drainage of an area since both infection and malignant tumours can spread by this route.

Consider:

● the **lymph vessels** that drain the area: superficial lymphatics tend to run with superficial veins, deep lymphatics tend to run with arteries (as with veins the division between superficial and deep is the deep fascia).

● the **primary receiving lymph nodes** and the extent to which these can be palpated.

● the **route** whereby lymph returns to the blood stream.

● the **degree of anastomosis**: in general there is very considerable anastomosis between the lymphatics serving adjacent areas so that, when lymphatics are blocked by tumour, nodes not normally draining an area may become involved. However, in the leg, there is relatively little anastomosis between the superficial and deep lymphatics.

Nerves

The central nervous system gives rise to 12 pairs of **cranial nerves**, and about 30 pairs of segmental **spinal nerves** (8 cervical, 12 thoracic, 5 lumbar, 5 sacral). **Peripheral nerves** contain **nerve fibres** of various types, together with their supporting **Schwann cells**, and delicate surrounding vascular connective tissue.

Somatic fibres supply the skin and musculo-skeletal system. **Autonomic** fibres supply the viscera, glands, and blood vessels. The autonomic system comprises **sympathetic, parasympath-**etic, and **enteric** subsystems; only sympathetic fibres enter the limbs. In all systems there are **motor** (efferent) and **sensory** (afferent) fibres. Cell bodies of somatic motor neurons lie in the ventral horn of grey matter in the spinal cord; their fibres are large and heavily myelinated. Smaller somatic motor neurons supply the contractile fibres in muscle spindles. Cell bodies of somatic sensory neurons lie in the dorsal root ganglia; their fibres range from large, heavily myelinated mechano-receptor fibres to small unmyelinated pain fibres. Cell bodies of sympathetic preganglionic neurons lie in the spinal cord between T1 and L2; their myelinated fibres are relatively short and end on sympathetic ganglion cells in the paravertebral chain of ganglia or the midline visceral ganglia. The ganglion cells give rise to unmyelinated postganglionic fibres that supply the blood vessels and viscera. Cell bodies of parasympathetic preganglionic neurons lie in the brain stem or in the sacral spinal cord; their myelinated fibres reach either one of the parasympathetic ganglia in the head or their target organ. Postganglionic parasympathetic neurons therefore have short fibres passing to their effector structures.

The cranial nerves supply parts of the head and neck and some viscera (Vols. 2, 3). Each mixed (motor and sensory) spinal nerve, on leaving the vertebral column, divides into an **anterior** and a **posterior primary ramus**. The posterior primary rami supply the extensor muscles of the spine and the skin over them; the anterior primary rami supply all the remainder, including the limbs. In the nerve plexuses that supply the limbs, the fibres of the anterior primary rami become segregated to form **anterior** and **posterior divisions** which supply, respectively, the muscles and skin of the (developmental) flexor and extensor compartments.

For any nerve that you study consider:

● the **types of fibre** that it contains.

In the limbs the somatic fibres may be motor, sensory, or mixed. The nerves also contain sympathetic fibres.

● the **origin** of the various fibres: for somatic nerves the spinal root, and peripheral nerve of origin; for autonomic nerves the ganglion or cranial nerve of origin.

● the **course** of the nerve, particularly where it is palpable or liable to trauma. Note that many sympathetic fibres travel as a plexus around major vessels.

● the **major branches** of the nerve.

● the **muscles supplied** by the nerve and the effect on movement of damage to the nerve.

● the **skin supplied** by the nerve and the area of anaesthesia produced by damage to the nerve.

● the **organ(s) supplied** by autonomic nerves, and their effects on the function of the organ(s). Remember that in the limbs sympathetic fibres are distributed not only to blood vessels, but also to sweat glands and erector pili muscles.

CHAPTER 3

Experimental examination of tissues

The **aim** of this chapter is to suggest some simple experiments which examine the function of living tissues.

The action of muscles

The actions of muscles can be investigated by many different methods: by performing different movements and feeling which muscles contract; by inferences from morphology and pulling on dissected tendons; by electrical stimulation; and by recording by electromyography, cinematography, flashing light photography, and kinesiology. Each method has its advantages and limitations. The results taken together provide more understanding than the results of any one of them taken alone.

Always remember that a muscle never acts in isolation, always as part of a number of groups of muscles involved in a movement. Also, remember that many muscles act over more than one joint. If such muscles are involved in producing movement at a single joint, their actions at the other joints over which they act must be prevented by the action of synergists (p. 9)

Physical examination

When studying any movement, feel your own body or that of a colleague to determine which muscles are contracting when the movement is performed against resistance.

Dissection and pulling on tendons

In the dissecting rooms you can observe the effects of pulling on tendons; combine this with inferences of the probable effect of shortening the distance between the origin and insertion of a muscle in relation to the possible movements of the joint(s) across which it pulls. However, since muscles do not normally act alone, this method shows nothing of the combinations of muscles which are naturally engaged in producing even the simplest reflex or voluntary movement.

Electrical stimulation of motor nerves to muscles

The nerves to superficial muscles can often be stimulated through the skin. This is safely done by use of an electronic stimulator which delivers 35 rectangular pulses per second, each pulse lasting 0.5 msec. The output voltage is controlled between 5–40 volts (see Appendix Fig. 1). A large stainless steel anode is covered with a pad soaked in

saturated NaCl solution and bandaged to a part of the body remote from the part being stimulated. A focal cathode of cotton wool soaked in saturated NaCl solution contained in and just protruding from a perspex tube is applied to the point of stimulation (**3.1**). The effective 'motor points' are at or near to the point of entry of motor nerves into muscles. Begin with low voltages and, using firm pressure, explore these points, increasing the voltage as necessary to elicit weak contractions. The effects of stimulating motor nerves to individual muscles can also be observed when these are exposed at operation.

Because no muscle normally acts alone, excessive electrical stimulation which causes isolated, powerful contraction of a muscle can cause dislocations. For example, powerful stimulation of deltoid can dislocate the shoulder! In natural abduction of the shoulder, the synergic actions of other muscles prevent this.

Recording muscle activity

Modern research on human movement involves recording the forces and movements at joints by force transducers, position transducers, and accelerometers ('kinesiology'), combined with simultaneous recording of the electrical activity of several of the muscles taking part (electromyography). Movements at several joints, or of the whole body, can be analysed by photographing flashing lights attached to the moving parts, or by cinematography, synchronized with electromyography.

The **electrical activity** of a muscle in any movement can be recorded. In clinical practice needle electrodes have to be used for deep muscles but you can record from a pair of superficial muscles (biceps and triceps are convenient) through the skin. The subject should recline comfortably on a couch in a warm room and practise muscle relaxation. Muscles to be investigated should first be identified by palpation during movements performed against resistance from the examiner's hand.

The areas of skin to which electrodes are to be attached (**3.2**) should be cleaned with 70% ethanol and allowed to dry. A large disc indifferent electrode is attached over the manubrium sterni, and small surface electrodes to either end of the belly of the muscle(s) being investigated. These electrodes can then be connected to either an amplifier and chart recorder, or to an amplifier connected to a loudspeaker (see Appendix, Fig. 2).

A possible series of investigations of biceps and triceps—both muscles have parts which cross more than one joint—can reveal a great deal about prime mover action, synergism, essential fixation, and

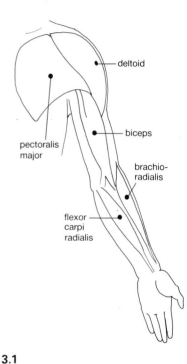

3.1
Suggested 'motor points' for stimulating with surface electrodes.

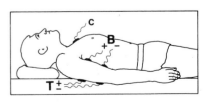

3.2
Suggested points for recording muscle electrical activity with surface electrodes. B biceps; T triceps; c indifferent electrode.

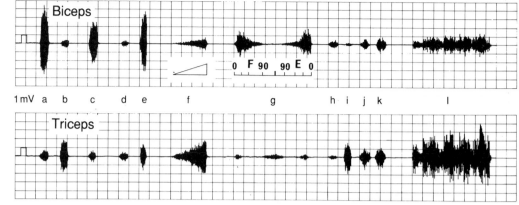

3.3
Record of muscle electrical activity from biceps and triceps during movements of the upper limb.
1 mV— calibration; see text.

action against gravity. We suggest that you investigate their electrical activity (**3.3**):

- during flexion and extension of the elbow against resistance with the forearm both supine (**3.3a,b**) and prone (flexion **3.3c**);

- during pronation (**3.3d**) and supination (**3.3e**) against resistance;

- during the gradual squeezing of an object (a dynamometer is ideal) in the hand (**3.3f**);

- during movement of the elbow from full extension to full flexion and back to full extension, with the distal end of the limb weighted to accentuate gravity (the forearm can conveniently be inserted in a tube with a 1 kg weight at the end) (**3.3g**).

- during flexion (**3.3h**), extension (**3.3i**), adduction (**3.3j**), and abduction (**3.3k**) of the shoulder joint.

There are, of course, numerous other possibilities; 'press-ups' (**3.3l**) give a very dramatic recording, often causing more triceps activity than can be produced by voluntary elbow extension. The aim of the experiments is to understand that a muscle has a role in many different movements, rather than the single (prime mover) 'action' usually quoted.

The skin

Sweat glands and dermal ridges

A replica (**3.4**) of the skin which shows dermal ridges and the number of actively–secreting sweat glands opening on to them can easily be made. Paint the area of skin with a solution of Formvar (4%) in ethylene dichloride, with Sudan black dye (1.5%) and dibutylphthalate (2%) (a plasticizer) added. When dry, the replica can be stripped off with adhesive tape, attached to a microscope side, and examined with a low-power (× 4) microscope objective. The differing distributions of sweat glands in a number of different areas (e.g. finger, dorsum of hand, forehead) can be examined. To demonstrate the effect of exercise make one replica, then exercise by running up and down stairs, and repeat the replica. You can then compare the number of open sweat glands in a standard area of replica.

3.4
Plastic cast of finger pad skin showing openings of active sweat glands onto dermal ridges.

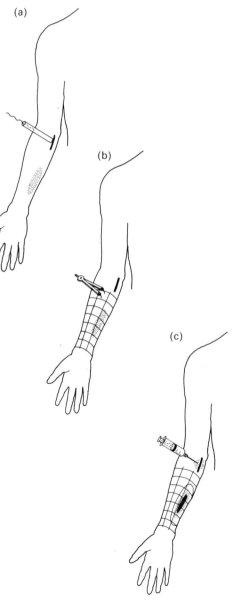

3.5
Investigation of sensory two-point discrimination on the skin. See text for details of procedure.

Skin innervation

You can test for the presence of cutaneous sense organs with different responses by exploring the skin (a) with a fine bristle; (b) with a warm (40°C) copper object and a cold (15–25°C) object; and (c) *gently* with a pin. Repeat your observations on the forehead and on the hand and note spatial variation in the threshold for touch sensation which is related, amongst other things, to the distribution of hair follicles. Cold 'spots' can be very precisely located because the receptors are very superficial, warm spots less so. You will probably find that you can locate separate spots on the palm of the hand but that, on the face, they are too densely packed to be separable.

Two-point discrimination (see p. 6) can be tested using an instrument such as blunted divider points. Test the discriminatory power of various parts, e.g. tongue, fingertip, palm, fore-arm, upper arm, back. The effects of anaesthesia on two-point discrimination can be tested. Locate and mark a cutaneous nerve which supplies an area of forearm skin by exploring along the known course of nerves with a unipolar repetitive electrical stimulus (see **3.5a** and Appendix, Fig. 1). Stimulation over a nerve produces a tingling sensation in a different, more distal area of skin. Mark squares 2×2 cm on to the forearm including in the centre of the grid the area supplied by the nerve (**3.5b**). Determine and record the variations in two-point threshold over the grid, preferably in both longitudinal and transverse axes because these may differ. Now ask a suitably qualified demonstrator to infiltrate some local anaesthetic around the marked nerve (**3.5c**). An area of skin within the grid should become totally anaesthetized within a few minutes. Determine and mark the boundary of the anaesthesia (if a small

nerve was infiltrated there may be no completely anaesthetized area). Now determine again the two-point threshold on the grid surrounding the anaesthetic area and also check that the anaesthetic area remains the same. Compare the two sets of data. You will find that the two-point threshold is diminished over a wider area of skin than that which has become anaesthetic. This is because cutaneous nerves supply overlapping territories.

The circulation

You should investigate on yourself or a colleague all the points at which arterial pulsations can be felt: similarly examine the course of the major superficial veins by lightly restricting venous return to an area. For the investigation of veins and venous valves see p. 69.

An ultrasonic Doppler-based probe can be used to map larger vessels. If one is available then choose an area such as the arm, forearm, or palm of the hand and map the course of the arteries and arterial arch (pp. 68, 72).

You should examine capillaries at the only two readily available sites. Clear the skin of the nail bed by application of a small amount of light oil and observe the capillary loops with a binocular dissecting microscope. By use of an ophthalmoscope, view the vessels on the surface of the retina through the pupil of the eye (see Vol. 3, p. 87).

If you or a colleague have an inflamed focus such as an infected finger or throat you may be able to feel enlarged, tender, and possibly painful lymph nodes, located respectively in the axilla and on the side of the neck just below the angle of the jaw, which drain these two areas.

CHAPTER 4

Medical imaging

One of the most important adjuncts to physical examination in the study of the anatomy of living subjects is **medical imaging**. Since the turn of the century images (**radiographs**) obtained with X-rays have been the standard method of visualizing many of the internal areas of the body. Images can be obtained by use of sound waves (**ultrasound imaging**), or by using radiation emitted from substances which have been administered to the patient (**nuclear imaging**). Recently, computerized techniques have been developed which display a cross-section 'slice' of the body. The first of these, **computed tomography** (CT), uses conventional X-rays and its principles have been extended to nuclear medicine to produce **emission computer-assisted tomography** (ECAT), and to the nuclear magnetic resonance effect to produce an image based on magnetic fields, now known as **magnetic resonance imaging** (MRI).

Radiology

In November 1895, when Wilhelm Roentgen was working on electron beams produced in a Crook's tube, he noticed light being emitted from material on a nearby bench. He realized that some form of emission other than electrons was passing through the tube and causing the material to fluoresce. Placing his hand between the tube and the fluorescing material he saw a shadow of his bones and realized the implications of his observation. By the end of the year he had published an important paper on the newly discovered 'X-rays' which rapidly led to the setting up of primitive X-ray departments in many hospitals.

X-rays are part of the electromagnetic radiation spectrum and can be produced by bombarding a tungsten anode with electrons using high voltages. When the electrons strike the anode their kinetic energy is converted to heat and radiation, including X-rays. In medical radiography the tungsten anode is suspended over the patient so that the beam of X-rays passes through the body; the emerging radiation is then picked up by a detector, which is usually photographic film (**4.2**). Since photographic film is sensitive to X-rays, the film will be exposed to a degree that depends on how much of the beam has passed through the patient and how much has been absorbed by the different tissues of the body. Air is radiolucent, bone and metal radio-opaque. In the chest, good contrast is provided by the bone, soft tissues, and air-containing lungs, and a clear image of most major structures such as heart (H), liver (L), and pulmonary vessels (arrows) can be obtained (**4.1**). In the abdomen, however, most

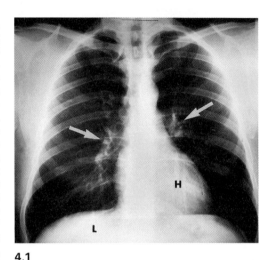

4.1
Radiograph of chest.

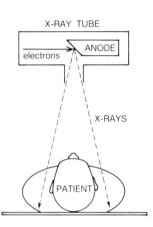

4.2
Conventional radiography.

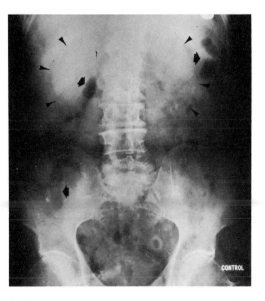

4.3
Radiograph of abdomen.

of the organs are of similar density with respect to X-rays and, although gas in the bowel (**4.3**, broad arrows) and bone can be distinguished, there is very little extra information to be obtained from soft tissue structures, although the outlines of the kidneys are just visible (**4.3**, arrowheads).

Fluoroscopy ('Screening')

If a fluorescent screen is substituted for photographic film a direct image is produced. This

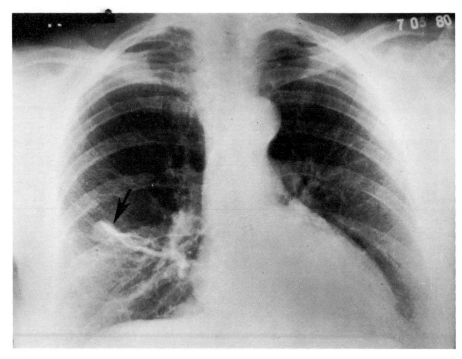

4.4
Radiograph of chest; opacity in right lung.

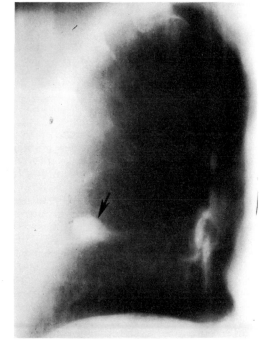

4.5
Tomogram of opacity seen in 4.4.

enables movements of organs to be studied. It is customary nowadays to view the image on a television screen; an amplifier system or 'image intensifier' is usually employed which ensures a good picture with a reduction in the dose of radiation. The fluoroscopic image can then be recorded by photography or video, or by a digital computer which stores the information received (digital radiography).

Tomography

Can be used to overcome the superimposition effects present in the conventional radiographic image. The X-ray tube and the film move around the patient in a constant relationship to the plane of interest, which therefore appears as a sharp image; the overlying and underlying areas move relative to the tube and film so that their image is blurred. The effect of tomography is therefore to produce a clear image of one plane only. It is widely used in the investigation of the chest, kidneys, and skeleton. Examine the conventional radiograph (**4.4**), which shows an opacity in the right lung (arrow). Note that in the tomogram (**4.5**) the plane of focus is such that only the nodular opacity is clearly seen.

Contrast media

It is often not possible to distinguish many organs by conventional radiography, especially in the abdomen. This problem can be overcome by the use of contrast media which are usually either pastes of inorganic barium salts for rectal or oral ingestion, or organic substances containing iodine for intravenous administration. To demonstrate the oesophagus and stomach, a suspension of

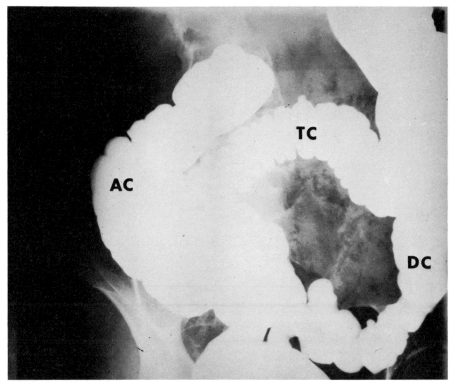

4.6
Radiograph of abdomen after barium enema.

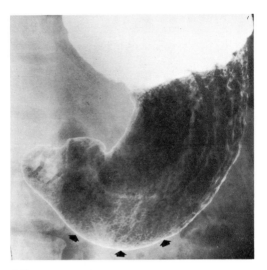

4.7
'Double contrast' radiograph of stomach. The patient has been tipped cranially causing the barium to collect in the fundus.

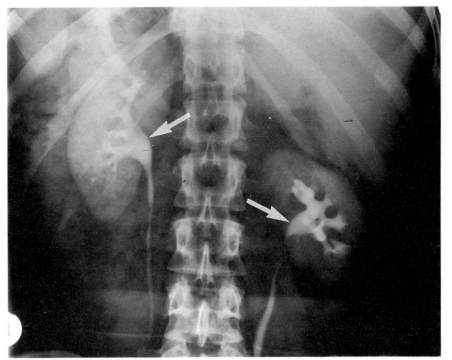

4.8
Intravenous urogram.

barium sulphate is given to a patient by mouth. The colon can be similarly studied if barium is introduced through the rectum; note that it has filled the descending colon (DC), transverse colon (TC), and ascending colon (AC) (**4.6**). More information is provided about the lining of the intestine if a small quantity of suspension is used and the gut is then distended with air, giving a 'double contrast' image. In **4.7** barium fills the upper part of the stomach of the recumbent patient, while the mixture of gas and barium allows detail of the lining mucosa of the lower part to be seen.

Iodine-containing contrast media have many applications; they can be injected intravenously to be excreted by the kidneys (**intravenous urogram**) (**4.8**). The contrast medium is concentrated by the renal parenchyma, making the kidney more obvious than in **4.3**, and is excreted into the renal collecting system, renal pelvis, and ureter (arrows). Injections of contrast medium can also be made into a joint (**arthrogram**; **7.2.15**), into the bronchial tree (**bronchogram**), or around the spinal cord (**myelogram**; **8.30**).

It is also possible to study blood vessels by inserting a fine tube (catheter) into an accessible vessel, for example the femoral artery, and passing its tip into the desired vessel, where contrast media can be injected (**4.9**). The catheter (arrow) has entered the left renal artery (arrow head) and contrast medium has outlined small vessels within the kidney. This procedure, called **angiography**, can be applied to vessels of the gut or head and neck, and even the chambers of the heart.

Ultrasound imaging

Sound waves travelling in a medium are partly reflected when they hit another medium of different consistency. This produces an echo and the time taken for the echo to reach the source of the sound indicates the distance from the reflecting surface.

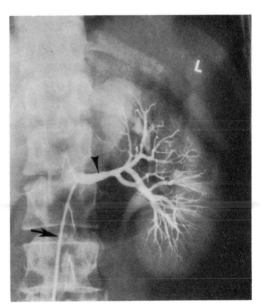

4.9
Arteriogram of left kidney.

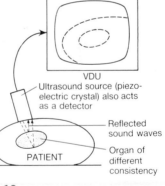

4.10
Ultrasound imaging.

Ultrasound imaging detects and analyses these echos. Ian Donald, Professor of Obstetrics in Glasgow, realized the potential of this technique and developed it to study the pregnant uterus and the fetus.

Ultrasound consists of high-frequency sound waves that cannot be detected by the human ear. When ultrasound travels in human tissue it undergoes partial reflection at tissue boundaries; thus a proportion of the sound waves return as an echo while the rest continue (**4.10**). Bone almost totally

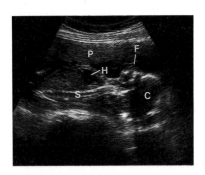

4.11
Ultrasound image of a pregnant uterus containing a 20-weeks-old fetus.

absorbs the sound and therefore no signal can be obtained from bone or the structures beyond. Neither can a signal be obtained from gas–containing viscera. These two facts are responsible for significant limitations to the use of this technique. However, it has the great advantage that there are no damaging effects at the sound energies used and it can therefore be used to monitor the developing fetus. In **4.11** the cranium (C), trunk, face (F), spine (S), heart (H), and placenta (P) of a 20-week-old fetus can be distinguished. The images are sectional in that they represent a 'slice' of the body in the plane of the ultrasound beam. In addition to its use in obstetrics, ultrasound is used in the investigation of the kidney, liver, and biliary system. Recently machines have been developed which have a rapidly repeated scanning action, enabling movement of various organs such as heart valves to be studied ('real-time' ultrasound).

Nuclear imaging

Nuclear medicine may be broadly defined as the use of radioactive isotopes ('radionuclides') in the diagnosis and treatment of disease. Interest to the student of anatomy arises from that branch of nuclear medicine called nuclear imaging or 'scintigraphy', whereby a 'map' of the uptake of radionuclide in a given organ system or pathological lesion can be produced by means of an instrument called a gamma camera. Analysis of such an image (which depends not only on morphology, but also on function) is then of use to the diagnostician in elucidating the nature of the patient's problem.

Historically, the subject began in 1896 with the discovery of radioactivity by Henri Becquerel, followed by the development of the radionuclide tracer method by Georg von Hevesy, the discovery of artificial radionuclides by Irene Curie and her husband, Frederic Joliot, and the invention of the gamma camera by Hal Anger.

The basis of the gamma camera is a large flat crystal of sodium iodide which converts into light rays the gamma rays emitted from radionuclides which have been injected into the patient. These light rays strike a photosensitive surface and cause the emission of electrons. The latter are amplified by a photomultiplier and eventually electrical pulses are formed. These electrical pulses can be used to produce an image on a television screen or be used as input to a computer system (**4.12**).

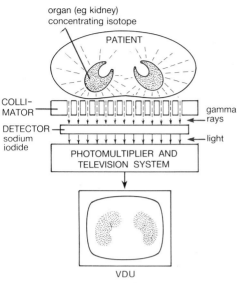

4.12
Nuclear imaging.

A number of radionuclides are used in a form described as 'radiopharmaceuticals'. The principal radionuclide used is Technetium–99m (which is very safe for the patient in terms of absorbed radiation dose, and has a convenient half-life of 6 hours), but others are also available, e.g. Iodine–123, Thallium–201, Gallium–67, Indium–111. These radionuclides are coupled to various compounds designed to deposit them in the organ of interest. For example, Technetium–99m coupled to sulphur colloid will be phagocytosed by macrophages and so be deposited in the reticuloendothelial system. Technetium–99m phosphates will deposit in bones. In **4.13** note the focal areas of increased isotope uptake in the spine and pelvis which are due to tumour deposits from breast cancer (arrows). Iodine–123 as sodium iodide will be taken up by the thyroid. The normal uptake of such a radiopharmaceutical in an organ must be well known to the radiologist, so that any deviation from the normal pattern is rapidly detected.

Sectional images (emission computed tomography) can also be produced using methods very similar to those for emission computed tomography (see below).

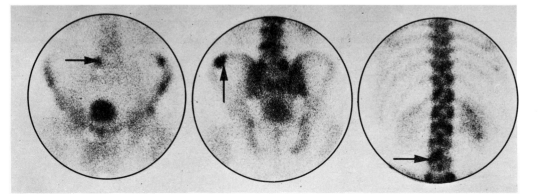

4.13
Nuclear images of pelvis and spine showing increased uptake of isotope consistent with tumour deposits.

Computed sectional imaging

Computed tomography

In the early 1970s Godfrey Hounsfield, of EMI Ltd, discovered a way of producing very clear cross-sectional images of the body using X-rays. This was hailed as the most significant advance in radiography since Roentgen's discovery and he, like Roentgen, the Curies, and von Hevesy, was awarded a Nobel Prize.

Computed tomography (CT) is similar to conventional radiography in that a beam of X-rays is passed through the body and is measured after it emerges. The differences between CT and conventional X-ray methods is that a very narrow beam is used and an array of highly sensitive photoelectric cells is substituted for photographic film. The beam is rotated around the patient and density measurements are made from many different angles. The data are analysed by a computer and the whole image is displayed on a TV monitor (**4.14**), dense structures such as bone conventionally being shown at the white end of the spectrum.

CT images are usually true axial sections although coronal sections can be taken of the head by positioning the patient appropriately. However, computer programs are available whereby a sagittal or coronal image can be built up from data derived from successive axial images. CT produces extremely clear and finely detailed cross-sectional radiographs without any superimposition of surrounding structures. Thus it is possible to see organs or small differences in density caused by disease which are almost impossible to demonstrate by conventional radiography. **4.15** is a transverse section CT of the abdomen in which the kidneys (K), liver (L), pancreas (P), and small bowel (B) are all clearly seen because they are outlined by body fat.

Emission computer-assisted tomography (ECAT)

This is based on the same principle as transmission computed tomography described above, except that it depends on gamma rays emitted from a radionuclide concentrated in an organ rather than on a transmitted X-ray beam (**4.16**). In **4.17** technetium-labelled methylene diphosphonate has been taken up in the skeleton of the chest in the same way as in the examination in **4.13**. This however is a cross-sectional image, so that you see the nuclide activity in the vertebral body posteriorly (V) and the sternum anteriorly (S). The uptake within the left side of the thorax (arrow) is in an area of muscle infarction in the heart. This would not have been seen on a conventional scintigram as it would have been obscured by activity in the overlying ribs.

Magnetic resonance imaging (MRI)

MRI is a new diagnostic technique based on radio signals emitted from resonating atoms within the body. The phenomenon of nuclear magnetic resonance has been known for some time and was developed by Sir Rex Richards, in Oxford, as a means of biochemical estimation of various substances. In the late 1960s this was extended to the analysis of human tissue and the first NMR image was produced by Paul Lauterbur of New York State University.

The nuclear magnetic resonance effect depends upon the fact that atoms carry a charge and can therefore be regarded as small magnets. When a large magnetic field is applied across a body there is a tendency for the nuclei to line up with this field. Nuclei spin and, rather like spinning tops, can be

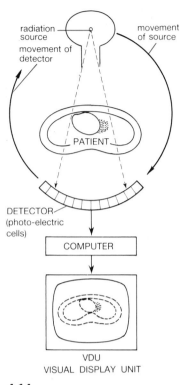

4.14
Computed tomography.

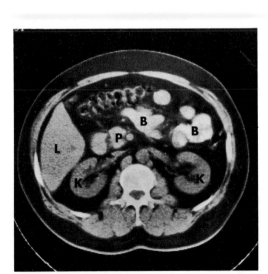

4.15
Transverse CT of abdomen.

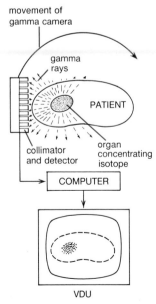

4.16
Emission computer-assisted tomography.

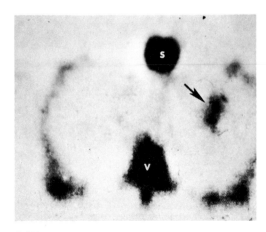

4.17
ECAT image of chest showing increased uptake of isotope in heart muscle consistent with tissue damage (infarction).

displaced from their axis by a field applied at an angle to the main one. If the second field is of an appropriate frequency the atoms can be made to resonate. Since the atomic nuclei carry a charge, any movement sets up radio signals which can be detected outside the body. The interest of magnetic resonance centres on the fact that the behaviour of the resonating nuclei depends upon the atoms surrounding them, in other words their chemical environment. Thus the radio signals measured outside the patient can reflect the biochemistry of the area under examination.

The magnetic resonance scanner looks rather like a CT machine in that the patient lies on a couch and enters a gantry which in this case is a very large electromagnet. Sections are taken through the area being examined and the resulting MR images are superficially similar to CT. However, CT is based on the **density of tissues** to X-rays whereas MR images reflect some aspects of the **biochemical composition of the tissues**. The sequence of magnetic pulses can be altered (T_1- and T_2-weighting) to vary the signal from different tissues but, in both, fat (including fatty bone marrow) produces a high signal intensity (white), muscle produces an intermediate signal, and cortical bone and fibrous tissue produce low signals (black). T_2-weighting produces a high-intensity signal from fluid, which gives a low-intensity signal with T_1-weighting. In **4.18** the section through the abdomen shows liver (L), spleen (S), stomach (St), and hepatic veins and ducts (arrows).

Digital radiography

In digital radiography each image is divided into a matrix of picture elements (pixels) and the density of each pixel is converted to a digit. The digitized image is stored on a computer and can be retrieved to a viewing screen, transmitted via telephone, and processed to improve its resolution. One method of processing is **digital subtraction angiography** (**4.19**) in which images of the background are subtracted from images of the same area after injection of contrast material. It provides more information than is possible with film, and enables images of the arterial system to be produced after intravenous injection of contrast material, which was not possible with conventional techniques. It also avoids the hazards of intra-arterial injection of contrast medium, and the expense of silver-coated film.

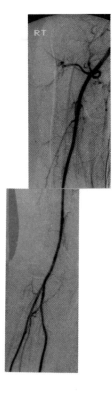

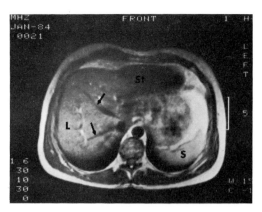

4.18
Magnetic resonance image of abdomen.

4.19
Digital subtraction angiogram of the main arteries of the arm and forearm.

Doppler colour-flow imaging

When a wave motion is radiated from a moving source there is a change in frequency of the wave—the Doppler effect. This principle can be used to study moving structures such as blood flowing through peripheral vessels. Probes emit ultrasonics of a given frequency. The signal received back from the body by the transducer contains the emitted frequency (due to waves reflected from stationary sources) and Doppler-shifted frequencies reflected from moving structures. The Doppler-shifted frequencies can then be filtered out, sorted for the extent of the shift, and recorded. The information can then be presented on a colour monitor with the vessels represented in different colours. An initial colour-flow image (**4.20a**) is used to locate the vessel being studied (in this case, the femoral artery) and, from the colour, to determine the direction of flow. Graphic presentation shows the speed of forward flow during systole and the small backward flow during diastole. This technique is often used to evaluate conditions in which flow is restricted (stenoses) or increased (tumour vessels).

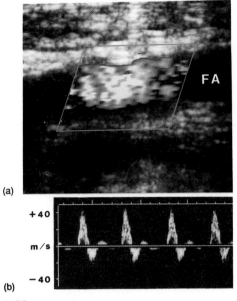

(a)

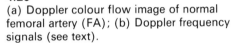

+40

m/s

−40

(b)

4.20
(a) Doppler colour flow image of normal femoral artery (FA); (b) Doppler frequency signals (see text).

CHAPTER 5

Embryonic development of the musculoskeletal system

Why study the embryonic development of the body? There are several reasons: embryology allows us to understand how, during development, the adult arrangements of tissues and cells are generated; it can help explain the aetiology (cause) of congenital anomalies; and it can reveal basic principles about cell behaviour that govern processes, both normal and pathological, such as cancer.

Some basic embryological concepts

Our bodies can be regarded as a **clone** of cells, all of which are derived from a single cell—the fertilized egg. All our cells contain identical genetic material (DNA), but there is great diversity in the characteristics of cells in different parts of the body; this is generated during embryonic development. For example, our musculoskeletal system is composed of characteristic tissue types, such as muscle, bones, cartilage, epidermis, dermis, blood vessels. The differences between the tissues must be established by the differential expression of particular, tissue-specific genes. Development must therefore involve the control of such specific gene expression.

But the complexity of our body does not cease at the level of the gross differences between cells that characterize different cell types. These cells are also arranged into characteristic **patterns**. The differences and similarities between our upper and lower limbs illustrate this well. Both contain the same cell types in roughly the same proportions, and yet they differ sufficiently that any first-year student can distinguish one from the other.

Thus, embryonic development can be viewed as consisting of two major processes: **differentiation**, by which cell diversity is generated; and **morphogenesis**—the generation of pattern. To understand development we must look at both processes. After the period when cell diversity and a gross morphological pattern are created, body parts undergo further refinements that generate the final, adult form. These include the rotation of the limb buds, which occurs differently in the upper and lower limbs, the consolidation of the bony elements (osteogenesis), and the formation of digits, which involves the death of cells lying between those that will form the fingers or toes.

Origin of the tissues that form the musculoskeletal system

All embryonic tissues are derived from three basic layers of cells that appear during the first two weeks of development. They are called **ectoderm** (a surface layer that gives rise to the epidermis of the skin and to the entire nervous system), **mesoderm** (which gives rise to most of the internal organs, such as muscles, dermis, circulatory system, etc.), and **endoderm** (which generates the epithelium of the gut and most of its associated organs).

Paired mesodermal structures called **somites** flank the embryonic axis (**5.1**, **5.2**). Each somite starts as a rosette which then loses its rosette structure and subdivides into two parts. The first is the ventromedial **sclerotome**, the cells of which come to surround the future spinal cord and give rise to the axial skeleton (vertebral column and ribs) (for the development of the spine see p. 141). Dorsolaterally, the somite remains an epithelium for longer, and this region is called the **dermomyotome**. From its most lateral part some cells become very active and invade the limb buds to generate skeletal muscles and probably also the dermis, whilst the dorsal part of the dermomyotome forms the **myotome**, which will give rise to the axial musculature.

Formation of the limb buds

The limbs are first visible in the human embryo as two pairs of **limb buds**, at about 4 weeks of gestation. By this time, the embryo measures about 4 mm in crown–rump length and already contains a variety of structures (**5.1**). The limb buds arise as paired evaginations (outpocketing) of ectoderm and the underlying lateral plate mesoderm. The ectoderm covering the limb bud gives rise to the epidermis of the skin; the mesoderm in the early bud gives rise to the bones, joint cartilage, and blood vessels of the limbs. The muscles and dermis form from mesoderm which migrates into the limb bud from the lateral part of the dermomyotome.

The innervation of the limb also originates outside the limbs. Cells in the ventrolateral part of the developing spinal cord (**neural tube**) produce axons which grow into the limb bud to innervate the muscles (motor nerves). Another population of

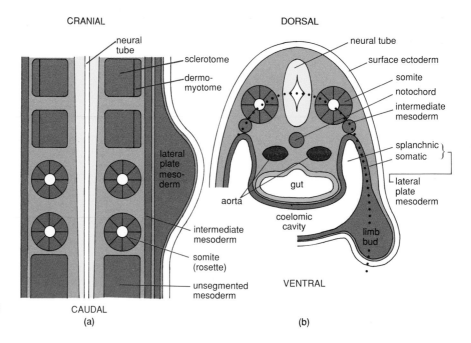

5.1
Diagrammatic (a) longitudinal section (along the dotted line in (b)) and (b) transverse section of an early embryo to show somites and other principal mesodermal structures. The left limb bud is not included.

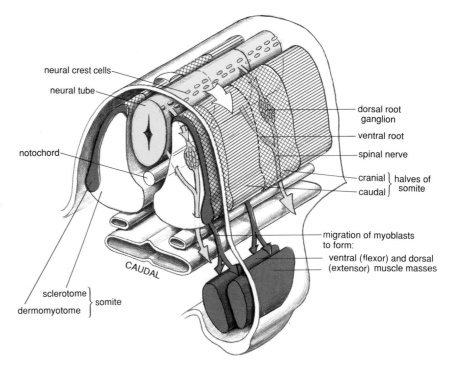

5.2
Diagram of early embryo. The myoblasts migrate from the dermomyotome into the limb where they form ventral (flexor) and dorsal (extensor) muscle masses. Motor innervation from the ventral root region of the neural tube migrates through the cranial half of each sclerotome where it joins sensory (dorsal root ganglion) nerve cells and their processes derived from neural crest cells. Other neural crest cells forming melanocytes migrate dorsal to the somites. Surface ectoderm forms skin, hairs, nails, and sweat and sebaceous glands.

neural tube-derived cells, which arises at the dorsal edge of the tube, is called the **neural crest**. These cells migrate very widely in the embryo and contribute to diverse structures which include the cranial vault and facial bones (Vol. 3), skin pigment, the sensory dorsal root ganglia, the whole of the autonomic nervous system, the adrenal medulla, and most of the Schwann cells (glial sheath cells) that surround peripheral nerves. On either side of the neural tube, neural crest cells develop into the dorsal root ganglia. Cells in these ganglia sprout axons that migrate both centrally into the dorsal part of the neural tube (to generate the dorsal root) and distally into the limb and other target regions to provide sensory innervation (**5.2**). Much of the evidence for this is derived from classic experiments in which neural crest cells from a chick embryo were replaced with neural crest cells from a quail embryo, which can be distinguished from those of the chick.

The development of pattern

It is important to realize that the pattern of the various elements in the limb (such as the shape of individual muscles, the shape and number of bony elements, etc.) is not a function of the type or origin of the cells in any particular part of the limb bud, but rather depends on the position at which cells find themselves with respect to other structures around them. Thus, if any two somites are exchanged by transplantation, either somite can give rise to the appropriate muscles and a normal limb develops. Experiments involving surgical removal or transplantation of specific regions can therefore be designed to find which, if any, portions of the limb bud are essential for establishing the pattern of these elements.

A developing limb has three axes, each of which apparently depends on a different mechanism to generate its pattern. The **proximo-distal axis** (shoulder to fingertips/hip to toe tips) seems to involve a mechanism that measures timing to specify different directions of development at different levels. The **pre-/post-axial axis** (radio-ulnar and tibio-fibular axes) appears to involve the action of a special signalling region which in the upper limb is located near the ulnar margin of the limb bud. The mechanism by which the **extensor–flexor axis** develops is very unclear.

Proximo-distal polarity. Recent evidence suggests that the specification of proximo-distal polarity depends on two special regions at the tip of the developing limb bud. At the tip of the bud there is a ridge, or ectodermal thickening, which follows the distal outline of the limb bud. This is called the **apical ectodermal ridge (5.3, 5.4)** and it is necessary for the continued outgrowth of the limb. If it is removed, the limb that develops is truncated

at a level related to the time of operation: the later the operation, the more distal are the last structures that form.

Immediately underlying the apical ectodermal ridge, in the limb bud mesenchyme, is a population of actively dividing cells called the **progress zone (5.3)** which supplies cells to the growing limb bud. The apical ectodermal ridge is required for the maintenance of the progress zone. If cell division at the progress zone is prevented, truncations of the limb pattern occur, equivalent to those seen after removal of the apical ectodermal ridge. Cells emerging from the progress zone appear to measure the time they spend in the zone and to use this information to determine their proximo-distal position, but how they do this is unknown. Congenital anomalies can affect this axis. For example, in phocomelia (a condition which can be caused by exposure of expectant mothers to the drug thalidomide at appropriate stages of gestation) distal limb elements (digits) develop prematurely and proximal elements are deleted. The digits are therefore connected directly with the shoulder.

Pre-/postaxial polarity. Most of what we know about limb development concerns this axis. Transplantation experiments in chick embryos **(5.4)** have defined a region near the post-axial (ulnar, in the upper limb) margin, which is capable of polarizing the entire pre-/post-axial dimension of the limb. This is called the **zone of polarizing activity.** If another zone is grafted pre-axially, all those limb elements developing after the transplantation become duplicated along this axis. Thus, mirror image duplications of all the digits can be generated by such a transplant (if the little finger is named '5' and the thumb '1', one might find the pattern: 543212345). Similar conditions occur spontaneously in humans, where they are called **poly-dactyly.** They may be due, at least in some cases,

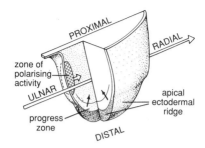

5.3
The limb grows from proximal to distal. The apical ectodermal ridge maintains the underlying progress zone. As cells leave the progress zone, the proximo-distal axis is determined. The zone of polarizing activity in the ulnar margin of the developing limb signals the radio-ulnar polarity.

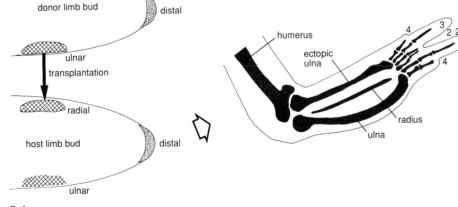

5.4
Experimental transplantation of a second zone of polarizing activity in the radial border of a developing chick limb results in the development of symmetrically duplicated distal limb skeletal elements.

to the presence of a second, ectopic, zone of polarizing activity at the pre-axial margin.

It is thought that the zone of polarizing activity emits a signalling molecule that diffuses freely across the width of the limb bud. The concentration of the molecule would inform cells in the limb bud mesenchyme of their distance from the post-axial margin. Vitamin-A (retinoic acid) or one of its derivatives may be directly involved in this signalling process, but we do not yet understand how.

Extensor–flexor polarity. We know very little about this axis, except the origin of the muscles that constitute the extensor and flexor compartments. As the prospective muscle cells derived from the dermomyotome of the somite (see above) approach the limb bud, they divide into two groups—**dorsal** and **ventral muscle masses**. The dorsal muscle mass gives rise to extensor muscles; the ventral mass produces the flexors. How more complex patterns are generated along this axis has yet to be discovered.

Refining the pattern

Formation of neural connections. The **motor innervation** of the limbs is derived from axons which grow out from the ventral root region of the neural tube (**5.2**). As they emerge, the axons become segmented into individual nerve roots because they are only able to grow through the cranial halves of the adjacent somites (**5.2**). As soon as the axons emerge from the somite, they splay out and sort out to form the limb plexuses. The nerves then grow toward the appropriate muscles, which they appear to recognize according to their position in the limb. In general, muscles developing near the radial side of the upper limb bud, such as biceps, are innervated by more cranially located motor nerves than those developing more caudally (such as triceps). Also, limb muscles are innervated in roughly a proximal-to-distal and cranial-to-caudal sequence. Thus, proximal muscles (e.g. shoulder, arm; C5–7) are innervated earlier and by more cranial motor roots than distal muscles (e.g. small muscles of the hand; C8–T1) (pp. 76, 131). Subsequent rotation of the limb buds (see below) partly obscures this arrangement.

Despite this apparently simple topographical correspondence, individual motor nerve roots are able to seek their appropriate muscles. If a small portion of developing neural tube is rotated about its cranio-caudal axis, axons emerging from the caudal (originally cranial) part of the rotated piece still innervate muscles located cranially in the limb (e.g. radial side) and those emerging cranially (original caudal) seek more caudally located targets (ulnar side). This experiment shows that individual motor nerve roots have defined identities which enable them to seek their appropriate target. The mechanisms underlying such remarkable behaviour are not understood.

The **sensory innervation** of the limb seems to obey similar principles. The dorsal root ganglia, which are derived from the neural crest, also develop in the cranial half of each somite. From there, they send sensory axons towards the limb to supply the nearest region of skin at the time of axon outgrowth. This is most easily visualized in the upper limb, where the sensory dermatomes[1] (**5.5** and see **6.9.3**) follow the limb outline in order: preaxial, distal axial, postaxial. In the lower limb, the same process occurs but the final arrangement is complicated by the rotation of the limb bud, which generates a spiral arrangement of the sensory dermatomes (see **7.8.2**).

Formation of digits; the role of cell death. During early stages of limb bud development, the distal part of the bud has a flat, palette-like shape. From this palette, the digits develop by two main processes. The pattern of skeletal elements of the digits is laid out in the mesenchyme inside the palette, by condensation of cells derived from the lateral plate mesoderm. At the same time, the regions intervening between the future digits undergo a process of 'programmed' cell death (in species such as ducks in which an interdigital membrane is found, the death of cells in this region is much diminished). Some forms of the human

[1] The term 'dermatome' has two quite different meanings. In Anatomy, it refers to a region of skin innervated by a single sensory nerve root. In Embryology, it refers to the portion of the dermomyotome of the somite that contributes cells to the dermis and to limb musculature. For this reason, the former are subsequently referred to as 'sensory dermatomes'.

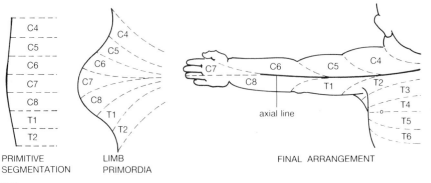

PRIMITIVE SEGMENTATION   LIMB PRIMORDIA   FINAL ARRANGEMENT

5.5
Development of segmental sensory innervation (dermatomes) of skin of upper limb.

congenital anomaly called **syndactyly** (fusion of fingers) may be due to defective patterns of cell death in the interdigital regions.

Development of limb vasculature

In each limb a primary axial artery develops in the mesenchyme and supplies a primitive capillary plexus. This main trunk in the upper limb forms from the fourth aortic arch and the sixth inter-segmental artery; it will form the definitive sub-clavian, axillary, and brachial arteries and, more distally, the anterior interosseous artery and deep palmar arch. In the lower limb the primary axial artery arises from the umbilical artery and follows the developing sciatic nerve. The external iliac artery and its continuation, the femoral artery, develop separately from the common iliac artery and provide a new channel to the lower limb which anastomoses with and eventually takes over the supply of the greater part of the lower limb.

Veins and lymphatics develop in the limb to form a deep system which lies alongside the major arteries and a more superficial system which lies in the subcutaneous tissue and which eventually drains into the deep vessels.

Chondrogenesis, osteogenesis, and the pattern of limb bones

The bones of the limb develop in the limb bud mesenchyme which condenses to form a 'model' of each bone. Chondrogenesis follows, during which the mesenchymal cells differentiate into cartilage. The process of osteogenesis then generates bone from these cartilaginous models (see Ch. 2 and Histology texts).

Rotation of the limbs

Later in development, the limb buds undergo complex rotations and, in the case of the lower limbs, a descent down the body axis. The upper limbs rotate only slightly, and therefore the sensory dermatomes still reflect the original organization of the sensory innervation. Nevertheless, the rotation is evident in the arrangement of the radius and ulna and the hand, which are rotated laterally by about 60° with respect to the humerus. In the lower limbs the rotation and descent are more marked, and the sensory dermatomes therefore assume a spiral pattern. The original sensory innervation of the lower limb buds with respect to the pre- and post-axial elements is retained, but the anterior surface of the thigh (extensor compartment) comes to face ventrally in the embryo and the (pre-axial) great toe comes to lie medially.

Rotation of the lower limbs is a relatively recent event during our evolution, and an adaptation to bipedalism. To visualize this process, think of how you would have to rotate your lower limbs to generate the position of the limbs in a lizard. The flexed knee would have its convexity facing dorsally and laterally. The reverse movement, back to the natural human position, mimics the rotation that occurs during human limb development.

CHAPTER 6

The upper limb: introduction

About 75 million years ago when primates appeared in the fossil record, one of the major adaptations to evolve was the increasing dominance of the forelimb for purposes of climbing, grasping, and feeding. Joints of the forelimb became more mobile, distal forelimb bones began to rotate about each other, and the thumb was set apart from the remaining digits to improve the grasp. With the emergence of hominids some 72 million years later bipedalism became increasingly successful and, with it, the 'emancipation of the forelimb'. Subsequent selection pressures led to the evolution of *Homo sapiens* with an upper limb set well away from the trunk by a long collar bone, and with the ability both to rotate the hand on the arm through 180° and also to oppose the thumb and forefinger.

These evolutionary adaptations have resulted in an upper limb which is the principal means whereby Man interacts mechanically with his environment. The upper limb is primarily the means whereby we position our hand to perform some function. The hand itself is specialized for grasping items and for moving them with various degrees of power and precision. The ridges which make up the fingerprint, and the firm anchorage of palmar skin both aid the mechanical interaction of grip. The hand can also be used to push or strike objects as can the forearm, arm, and shoulder. The more proximal parts of the limb, which contain relatively coarse muscle groups in comparison to the small muscles responsible for exquisitely precise finger movements, can therefore function both in their own right, and to steer and position the hand. The shoulder joint is a multiaxial ball and socket joint and the shoulder girdle itself is mobile, being attached to the trunk largely by muscle. As a result, the humerus, and thus the hand, can be moved widely in any direction with respect to the trunk. The elbow is a hinge joint; but its movement, combined with the pronation and supination movements of the forearm and the movements of the wrist, enables the hand to be moved widely with respect to the humerus. In the hand, all the digits, but particularly the thumb, can be moved independently; movements of the more proximal bones enable the palm to take up particular configurations during grasping movements. These developments in mechanical function could not have taken place in isolation and without parallel changes in the central and peripheral nervous systems which control the various groups of muscles involved in any movement.

The hand is also one of the principal means by which Man actively explores his environment. The skin of the fingers is therefore particularly richly supplied with sensory receptors which convey mechanical, thermal, and painful stimuli to the central nervous system for analysis and evaluation. Movements of the fingers and hand around an object enables us to identify complex shapes and the materials of which they are made without looking at them; the sensory system also provides essential feedback in the control of fine movements.

It is important, therefore, that while studying the upper limb, its role in communication via sensory information and gesturing, in balance, in carrying, in making a variety of different grips, and in performing delicate manipulations, should all be carefully considered.

Skin of the upper limb

Before starting to study the internal anatomy of the upper limb, review the section on **skin** in Ch. 2. The skin is the interface between body tissues and the outside world. It acts to protect the body from injury and invasion from without and from loss of fluid from within; loss of 40% of body skin (e.g. by burning) is usually fatal. Skin is also the mechanical interface through which forces generated by our muscles can act on objects around us, and a sensory interface through which we analyse our contact with our surroundings.

Examine the skin covering your upper limb. Note its varied appearance in different areas. The extensor surface is more **hairy** than the flexor surface and the palm is hairless. Except in the palm, the **texture** of the skin is coarser over the extensor aspect (especially at the elbow) than over the flexor aspect. Look at the skin around joints at which flexion and extension occur (e.g. elbow, wrist, fingers) and locate the distinct **skin creases** which mark firm attachment of the skin to the underlying deep tissues. Examine the skin creases on your hand as you move your thumb, index finger, and 3rd–5th fingers; the creases reflect the independent movement of the thumb and index finger. In Down syndrome, there is usually a single 'simian' transverse crease across the palm at the base of the fingers. Examine the fine **dermal ridges** on the skin that form the 'fingerprint' and 'palmprint' and compare them with those of your partner. By use of a hand lens, locate the openings of eccrine **sweat glands** on the forearm and on the dermal ridges (see **3.4**). These glands are derived from the epidermis of the skin and secrete a watery saline which, in evaporating, draws heat from the skin and thereby assists in thermoregulation. Blood flow through the skin is also important in thermoregulation; to assess this look at skin colour, especially the nails and finger pads. Watch one finger pad carefully and prick an adjacent finger pad with a pin. The sweat glands of the fingers, unlike those of the rest of the body, respond to alerting or emotional rather than to thermal stimuli. Pain and temperature are only two of the cutaneous sensations; our fingers in particular are immensely sensitive to touch and vibration.

Examine the **axilla** (armpit) and note the coarse texture of its hair which starts to grow at puberty under the influence of androgen hormones. Many axillary sweat glands are apocrine in type; their secretion contains remnants of cells lining the gland lumen. Breakdown of this organic material by skin bacteria causes the formation of odours which are said to act as sexual attractants.

Seminar 1

Bones of the upper limb

Aims To study the skeletal framework of the shoulder girdle and upper limb in the living body. To study the individual bones, including their surface contours and internal architecture, and to understand how these reflect the forces and articulations to which the bones are subject; to study their growth and development; to consider their skeletal functions including that of calcium storage; and to recognize that all bones are 'alive' and constantly, if slowly, renewing their form.

A. Living anatomy and the bony skeleton

For this and other seminars in which you will be examining your own body and that of a partner it is essential that you wear clothes that will permit such an examination (e.g. a sports vest). Note your findings alongside the text and figures in this book for rapid recall at a later date. Before you start the work of this seminar, review the section on bones in Chapter 2 (p. 6).

Identify on your partner and on an articulated skeleton the bony features of the shoulder girdle and upper limb illustrated in **6.1.1–6.1.3**.

Bones of the shoulder girdle

Clavicle

- sternal end
- acromial end
- the sigmoid shape of the bone
- roughened areas on the inferior surface

The clavicle acts as a strut holding the upper limb, to which it is attached, away from the trunk. It also transmits forces from the upper limb to the axial skeleton. The clavicle is unusual in that it is the only 'long' bone to lack a medullary cavity.

Pick up a clavicle and note the roughened areas on its under surface, at either end and in the middle.

Qu. 1A *What forces might have been responsible for these bony roughenings during the course of development?*

Qu. 1B *Orientate the isolated clavicle and place it as close as possible in its correct position on your partner. How can you tell which is a right or left clavicle?*

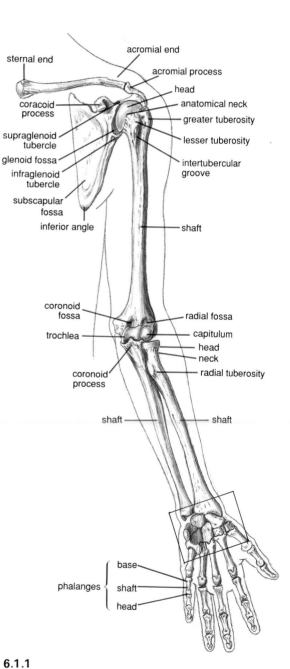

6.1.1
Bones of upper limb and shoulder girdle; anterior view. (For detail of carpus see **6.1.3**.)

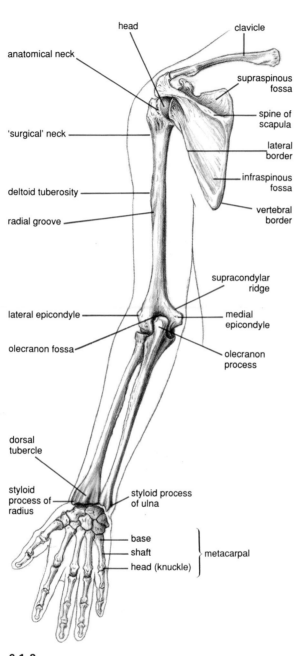

6.1.2
Bones of upper limb and shoulder girdle; posterior view.

Scapula

- coracoid process
- acromial process
- spine of the scapula
- supraspinous fossa
- infraspinous fossa
- subscapular fossa
- lateral and vertebral borders
- inferior angle
- glenoid (articular) fossa
- supraglenoid tubercle
- infraglenoid tubercle

Place a scapula as close as possible to its correct position against your partner. Now repeat this with your arm raised above your head and note the change in its position.

Bones of the upper limb

Humerus

- head, anatomical neck, and shaft
- surgical neck (a common site of fracture)
- greater tuberosity
- lesser tuberosity
- intertubercular (bicipital) groove
- deltoid tuberosity
- lateral epicondyle
- medial epicondyle
- trochlea
- capitulum
- olecranon fossa
- radial fossa
- coronoid fossa
- radial (spiral) groove

Ulna

- shaft of the bone
- coronoid process
- olecranon process
- trochlear notch (not labelled)
- radial notch for articulation with head of radius (not labelled)
- distal end (head) of ulna with its styloid process

Radius

- head (proximal end), neck, and shaft
- radial tuberosity (bicipital tuberosity)
- dorsal tubercle on the distal end of the radius (Lister's tubercle)
- expanded distal end of radius with its styloid process and facet for the ulna (ulnar notch)

With the upper limb held straight by the side and the palm facing forwards you will note that the forearm is angled laterally with respect to the arm; this is known as the 'carrying angle'. It is more pronounced in women than in men.

Carpus

The carpus as a whole is arched transversely owing to the shape and articulation of the carpal bones.

- Scaphoid (S), its tubercle (t), waist, and proximal pole
- Lunate (L) } proximal row
- Triquetral (T)
- Pisiform (a sesamoid bone) (P)

- Trapezium (Tm), its groove (g) and ridge
- Trapezoid (Td)–distal row } distal row
- Capitate (C)
- Hamate (H), and its hook (h)

Qu. 1C *Which carpal bone would transmit the greatest force if a person fell on to the outstretched hand?*

Metacarpus

Each of the 5 metacarpal bones has a
- base (that of the third is particularly prominent dorsally)
- shaft
- head (knuckle)

Phalanges: proximal, middle, and distal

- bases
- shafts
- heads (note the expansions on distal phalanges which support the pads of the fingers)

B. Radiology

When radiographs are examined the following three points should be routinely checked.

i) The outline of the bone, which includes the various prominences already mentioned, is important since the diagnosis of an abnormal appearance depends on a clear knowledge of that which is normal.

Qu. 1D *Compare the radiographs (6.1.4, 6.1.5) and comment on their appearance; which is abnormal and why?*

ii) The consistency of the bone; note the appearance of the hard outer cortical bone and the less opaque inner cancellous bone, remembering that their relative thickness differs in different bones and in different parts of the same bone. Sometimes the fine bony lines (trabeculae) in the cancellous areas have a characteristic pattern.

Qu. 1E *What abnormality can be seen in 6.1.6?*

iii) The relationship between the ends of bones; where they meet to form joints (this aspect will be dealt with in later seminars).
Examine radiographs **6.1.5**, **6.1.7–6.1.9**, **6.1.13** and see **6.4.2**, **6.4.3** and identify all the features that you have identified on the isolated bones.

C. The development of upper limb bones

Review the section on the ossification of bones in Chapter 2 (p. 6). All the upper limb bones develop by ossification in cartilaginous models except for the clavicle.

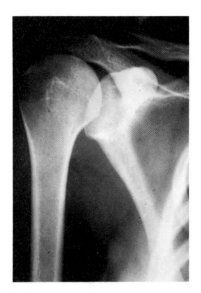

6.1.5
Radiograph of shoulder joint (adult).

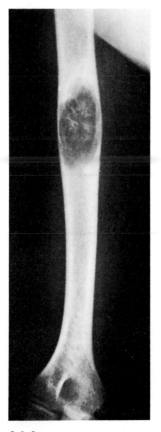

6.1.6
Shaft of humerus; Qu. 1E.

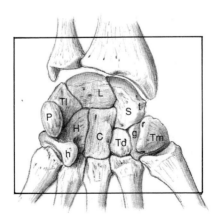

6.1.3
Carpal bones; anterior view.

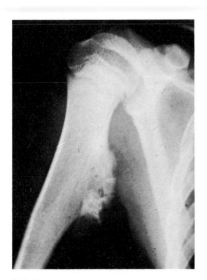

6.1.4
See Qu. 1D.

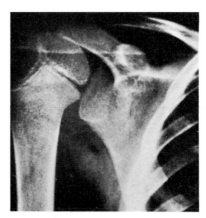

6.1.7
Shoulder (AP; 15 yrs).

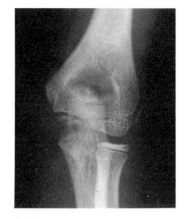

6.1.8
Elbow (AP; 15 yrs).

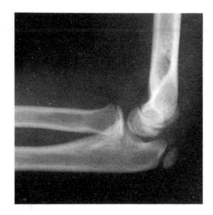

6.1.9
Elbow (lateral; 15 yrs).

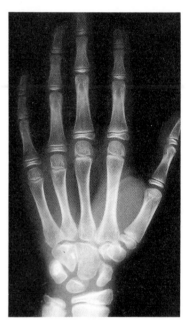

6.1.10
Hand (AP; 15 yrs).

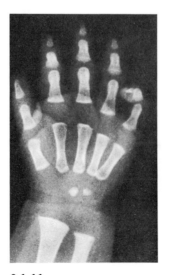

6.1.11
Hand (2 yrs).

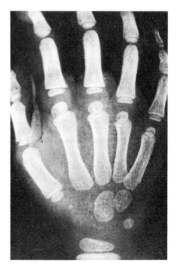

6.1.12
Hand (4 yrs).

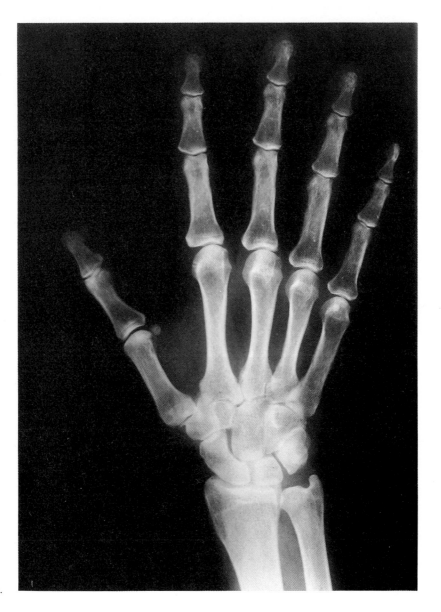

6.1.13
Hand (adult).

The clavicle is the first bone in the body to ossify. This starts at about the 6th week of intrauterine life at primary centres which ossify a membranous model of the shaft of the bone. Late in puberty, secondary centres appear in a small area of cartilage at the sternal and sometimes the acromial end of the developing bone.

All the other 'long' bones of the upper limb, including the metacarpals and phalanges, are formed by ossification of a cartilaginous model. Primary ossification centres appear in their shafts at about 8 weeks of intrauterine life. Secondary centres at the ends of the bones appear after birth and fuse with the shafts as linear growth ceases at the end of puberty. The proximal end of the humerus and the distal ends of the radius and ulna are the 'growing ends'. The carpal bones ossify from single centres which form during childhood.

Examine the radiographs of the upper limb and note the secondary centres of ossification, especially those of the elbow region—fractures of the elbow in children are relatively common. Unless one is aware of the pattern of ossification it is possible to mistake an epiphyseal plate for a fracture.

Identify the bones and epiphyses in **6.1.7–6.1.12**. Note especially the scaphoid and the lunate in **6.1.13** and the overlapping articulation of the distal row of carpal bones with the bases of the metacarpals. These overlapping surfaces can also simulate fractures unless they are carefully delineated.

Qu. 1F *Note the secondary centres of ossification in the fingers. The first metacarpal (of the thumb) might be referred to as a phalanx; why?*

Requirements:

An articulated skeleton
Separate bones of the shoulder girdle, arm, forearm, and hand
Radiographs of the upper limb at different stages of development.

Seminar 2

Attachment of the upper limb to the trunk

Aims To study the various means by which the upper limb is firmly yet flexibly attached to the trunk; and to understand how, through the co-ordination of movements between the chest wall, the scapula, and the humerus, a remarkably wide range of movements of the upper limb can be achieved.

A. Living anatomy

Ensure that you are dressed so that the shoulder girdle can be examined. Review the sections of Chapter 2 on joints and muscles.

Feel the outline of the **clavicle** on your partner, noting that its anterior aspect is convex medially and concave laterally. The clavicle acts as a strut, holding the shoulder joint away from the chest wall. This permits a considerable degree of mobility of the upper limb. The lateral end of the clavicle articulates with the acromion of the scapula at the **acromioclavicular joint**; its medial end with the manubrium of the sternum at the **sterno-clavicular joint**, the only joint attaching the shoulder girdle to the trunk. The upper limb and shoulder girdle are essentially slung from the skull and axial skeleton by muscles.

Fractures of the clavicle occur mostly in the middle third of the bone and are due to falls on the shoulder or outstretched arm when the weight of the body is transmitted along the shaft of the bone, which is relatively weak at the junction of its two curves.

The **sternoclavicular joint (6.2.1)** is a syno-vial joint of the plane variety and can therefore glide in all directions. Feel your partner's joint and get him to raise (elevate) and lower (depress) his shoulder and to move it forwards (protract) and backwards (retract). Record the movements that you feel at the sternoclavicular joint. Repeat this on yourself.

The **acromioclavicular joint** is also a syno-vial joint of the plane variety. Palpate this joint on your partner during movements of the shoulder girdle.

Examine both joints on an articulated skeleton. Note that these joints, with the clavicle, transmit forces applied to the scapula (from the arm) to the axial skeleton.

Next examine the movements of the scapula while palpating its inferior angle and spine. The scapula is retracted as the shoulders are braced back, and is protracted around the chest wall as the arm pushes forwards. The scapula can be elevated

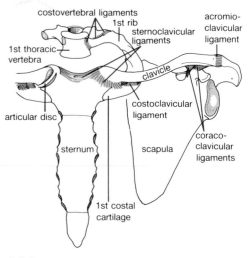

6.2.1
Bones and ligaments of shoulder girdle.

or depressed as the shoulders are raised or lowered. Examine the **rotation** of the scapula that occurs as the arm is raised above the head. Try to repeat the movement while preventing the scapula from moving by grasping its inferior angle; the degree of arm abduction possible will be less than 90°.

Instructions for demonstrating the action of the living muscles are included with the study of the muscles in prosections.

B. Prosections

Attachment of the clavicle and scapula to the torso (6.2.1)

If you are dissecting rather than using prosections, the skin incisions you should make are shown in **6.2.2**.

On a prosection of the **sternoclavicular joint** examine:

● The fibrous capsule which unites the two bones, and its lining of synovial membrane.

● The **superior sternoclavicular ligament** which strengthens the capsule superiorly (it is also known as the interclavicular ligament because it extends across the midline to the joint on the opposite side); and the anterior and posterior ligaments which also strengthen the capsule.

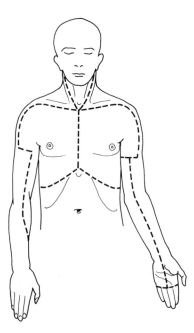

6.2.2
Skin incision lines.

● The fibrocartilaginous **intra-articular disc**: which is attached around its margin to the capsule. Its lower part is tucked under the medial end of the clavicle to be attached to the first costal cartilage.

● **Subclavius**—a muscle which attaches the under surface of the clavicle to the first costal cartilage, pulling the clavicle into the sternoclavicular joint and stabilizing the joint.

● The short, tough **costoclavicular ligament** which attaches the sternal end of the clavicle to the first costal cartilage.

● The very strong **coracoclavicular ligaments** which attach the under surface of the distal part of the clavicle to the coracoid process of the scapula.

Examine also a dissection of the acromioclavicular joint and the thickenings of its capsule.

Qu. 2A *Which of the two bones will lie uppermost in a dislocated acromioclavicular joint (6.2.3)?*

Qu. 2B *The clavicle is a commonly fractured bone (6.2.4). During what sort of accident would it be subjected to a major compression stress along its length?*

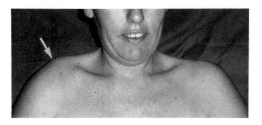

6.2.3
Dislocated right acromioclavicular joint (arrow); compare with the patient's left side.

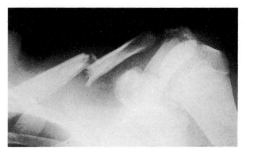

6.2.4
See Qu. 2C.

Muscles of the back

Examine the attachments of the upper fibres of **trapezius** (**6.2.5**). They arise not only from the external occipital protuberance of the skull and a bony ridge which runs laterally from it (superior nuchal line), but also from the spines of the cervical vertebrae via the thick nuchal ligament. The fibres are inserted into the inner aspects of the lateral end of the clavicle and the acromial process and the upper aspect of the spine of the scapula.

Face your partner and, placing your palms on both his shoulders, get him to shrug; feel the contraction of the upper fibres of trapezius.

The middle fibres of trapezius arise from the spinous processes of the upper thoracic vertebrae and are inserted into the upper aspect of the spine of the scapula. To feel these fibres contracting, ask your partner to brace back his shoulders.

The lower fibres of the muscle arise from the spinous processes of the lower thoracic vertebrae and pass upwards to be inserted into the apex of the smooth triangular surface at the medial end of the spine of the scapula. Because of the direction of the latter fibres, one of the actions of the muscle is to help rotate the scapula on the chest wall so that the glenoid fossa of the scapula faces upwards. Ask your partner to raise his arm straight above his head and feel the rotation of his scapula as he does so.

The weight of the upper limb is transferred to the trunk mainly via the coracoclavicular ligament (**6.2.1**) and by trapezius.

Qu. 2C *6.2.4 is a radiograph of a fractured clavicle; What forces would cause the two ends of the bone to take up the positions in which they lie?*

Examine the prosected part and turn the trapezius laterally thereby exposing the muscles **rhomboid major** and **minor**, which attach the vertebral border of the scapula to the spines of the upper thoracic vertebrae. Inserted into the superomedial angle of the scapula find **levator scapulae**, a muscle which arises from the posterior aspects of the transverse processes of the upper four cervical vertebrae.

Qu. 2D *What is the action of the rhomboid muscles?*

Examine **latissimus dorsi** (**6.2.5**) which arises from thick lumbar fascia attached to the spines of the lumbar and sacral vertebrae, the posterior third of the iliac crest, and the spines of the lower six thoracic vertebrae. The muscle passes upward over the angle of the scapula (to which a bundle of muscle fibres is often attached) to be inserted, via an aponeurosis, into the intertubercular groove of the humerus. With the arm at the side the aponeurosis, like that of pectoralis major, has a spiral arrange-ment. When the arm is raised above the head the spiral is straightened.

Qu. 2E *What is the action of latissimus dorsi?*

Muscles of the anterior and lateral aspect of the thorax

Examine the attachments of **pectoralis major** (**6.2.6**, **6.2.7**). It has two heads which arise, respectively, from the side of the body of the sternum and the upper six costal cartilages, and from the anterior aspect of the medial third of the clavicle. Its fibres spiral to be inserted into the lateral border of the intertubercular groove of the humerus.

Qu. 2F *What are the actions of pectoralis major?*

Examine **pectoralis minor** (**6.2.7**) which lies beneath pectoralis major. It arises from the anterior aspect of the 3rd, 4th, and 5th ribs and inserts into the tip of the coracoid process of the scapula.

Qu. 2G *What is the action of pectoralis minor?*

Ask your partner to bend his head forward against resistance from your hand on his forehead. On either side of the neck note the contraction of **sternomastoid** (sternocleidomastoid), which arises from the manubrium of the sternum and the sternal end of the clavicle; it inserts into the mastoid process of the temporal bone (a bony protuberance of the skull immediately behind the ear) and into the occipital bone behind.

Qu. 2H *How can you get the sternomastoid of one side only to contract? What movement takes place if both muscles contract?*

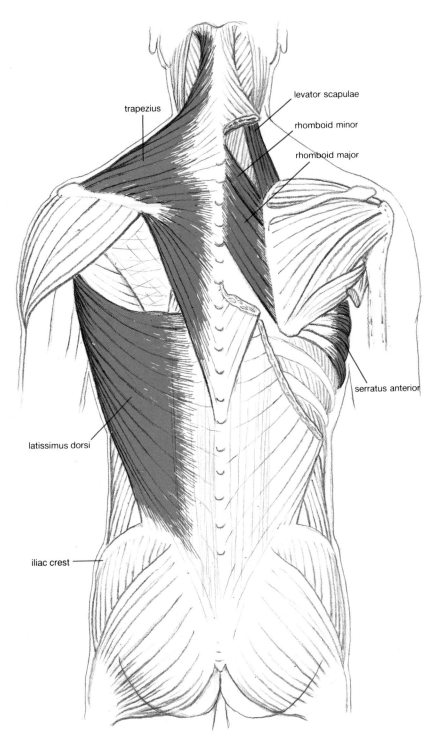

6.2.5
Superficial muscles of back and shoulder girdle.

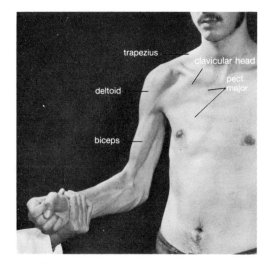

6.2.6
Anterior muscles of arm and shoulder girdle demonstrated by adduction and medial rotation of the arm against resistance.

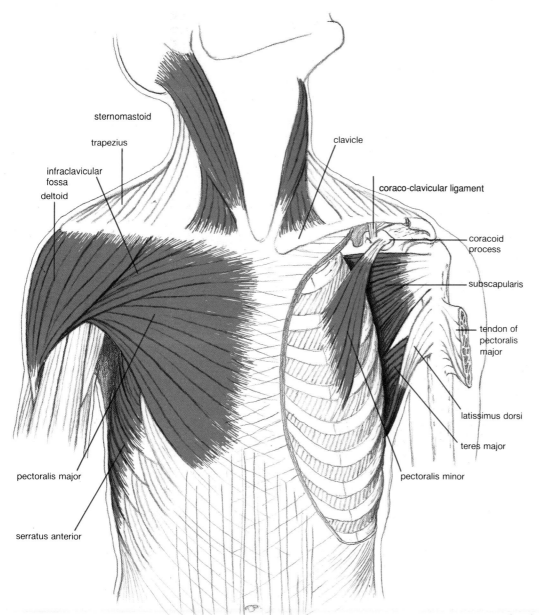

6.2.7
Muscles of axilla; pectoralis major removed on left side to show pectoralis minor and muscles of posterior axillary wall.

Examine **serratus anterior** (**6.2.7, 6.2.8**) which arises by individual slips of muscle (digitations) from each of the upper eight ribs; the slips unite and pass backwards around the lateral wall of the chest to insert into the whole length of the vertebral border of the scapula on its costal aspect. With pectoralis minor it protracts the scapula.

The axilla

Place the flat of your hand on serratus anterior and slide your fingers upward beneath pectoralis major and minor into the pyramidal 'space' called the

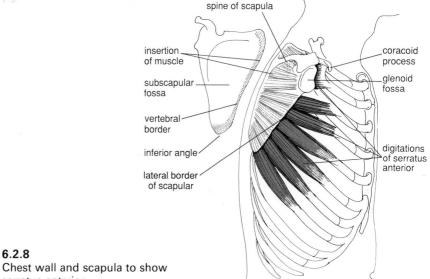

6.2.8
Chest wall and scapula to show serratus anterior.

axilla. Note that pectoralis major and minor lie in front of your hand, forming the **anterior wall** of the axilla (**6.2.7**). Subscapularis and teres major (p. 44), with the aponeurosis of latissimus dorsi spiralling around it, form the **posterior wall**. The **medial wall** of the axilla is formed by serratus anterior with its nerve, lying against the upper part of the thoracic cage.

Neural elements of the brachial plexus arising in the neck pass downward between the clavicle (anteriorly) and scapula (posteriorly); at the first rib they are joined by the first thoracic nerve root, and the subclavian (later termed axillary) artery and vein, all of which emerge from the thorax. The entire **neurovascular bundle**, ensheathed in fascia continuous with that of the neck, crosses the first rib to enter the **apex** of the axilla where it is further surrounded and protected by fat. Within the axillary fat are groups of lymph nodes which drain the upper limb and limb girdle, the superficial aspects (skin and superficial fascia) of the front and back of the chest wall, and the breast (see p. 88).

Grip, in turn, the lower border of the anterior and posterior walls of your own axilla. The **floor** of the axilla is dome-shaped because suspensory bundles of fascia continuous with that covering pectoralis minor are inserted into the skin and superficial fascia of the axillary floor.

When you have completed Seminars 9–12 on the innervation of the upper limb, make a note of the nerves which supply the muscles you have been studying.

Requirements:

An articulated skeleton
Prosections of the sternoclavicular joint; coraco-clavicular ligaments; superficial and deep muscles of the back; muscles of the anterior and lateral aspects of the chest
Radiographs of acromioclavicular and sterno-clavicular joints; fractured clavicle.

Seminar 3

Shoulder joint

Aim To study the wide range of movement at the gleno-humeral joint; to consider structural adaptations of this synovial 'ball and socket' joint for both movement and stability; and to study the muscles responsible for its movements.

A. Living anatomy

The functional axes of the shoulder joint are determined by the plane of the scapula as it glides around the chest wall—i.e. about 45° forward from the transverse axis (see **6.3.6**). Explore the range of possible movements at the shoulder joints on yourself and your partner: flexion and extension; abduction and adduction; medial and lateral rotation; and circumduction—a combination of all these movements. Perform all these movements again, at the same time palpating the anterior aspect of the head of the humerus. Make certain that you can feel the greater and lesser tubercles of the humerus and the intertubercular groove as you rotate the arm medially and then laterally.

Place your right upper limb in the 'anatomical position (p. 2) and abduct your shoulder as far as possible without rotating the humerus; the abduction will be limited to about 90°. Now raise your arm above your head naturally and note the lateral rotation of the humerus that occurs. Perform the same movements on an articulated skeleton and note that the rotation makes more articular surface available for the abduction movement.

Qu. 3A *When movements are occurring at the shoulder joint are movements also occurring at other joints?*

Qu. 3B *Can all the movements at the shoulder joint be performed if the scapula remains fixed?* (Review the associated scapular movements; p. 38.)

The contour of a normal shoulder, when looked at from in front, is formed from above downwards by the trapezius, the acromion and the acromio-clavicular joint, and the deltoid overlying the greater tubercle of the humerus. The deltoid also has a forward convexity due to the underlying greater and lesser tubercles of the humerus. Medial to this is a depression beneath the clavicle, the infraclavicular fossa, in which the coracoid process of the scapula can be felt.

B. Radiology

Examine radiographs which show the shoulder joint in various positions. Note that the articular surfaces of the humeral head and the glenoid fossa lie parallel to one another (see **6.1.5**). Confirm the rotation of the humerus during abduction from the positions of the greater and lesser tubercles of the humerus in full adduction and full abduction. See also the MR images of the shoulder **6.3.6**.

C. Prosections

The shoulder joint

Examine a prosected shoulder, noting the features that are adaptations for mobility. Identify the shallow **glenoid fossa of the scapula** which is covered by hyaline cartilage and deepened around its rim by a **glenoid labrum** of articular fibrocartilage. Note that only a portion of the almost spherical **head of the humerus** forms the articular surface.

Examine the **capsule** (**6.3.1**) which is attached around the cartilaginous rim of the glenoid fossa and to the anatomical neck of the humerus except inferiorly where the attachment passes downwards on to the medial aspect of the shaft. The capsule is loose inferiorly. This, with the shallow articular surfaces, permits a wide range of movement. Note the glistening **synovial membrane** which, in all synovial joints, lines the capsule and all those parts of the joint which are not covered by hyaline

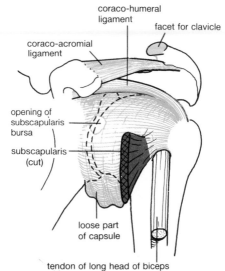

6.3.1
Shoulder joint: attachments of the capsule and extrinsic ligaments.

articular cartilage. Its function is to produce **synovial fluid**.

Qu. 3C *What are the functions of synovial fluid? What is meant by its thixotropic properties?*

Examine the rather insubstantial thickenings or **intrinsic ligaments** of the capsule. These can be found anteriorly (glenohumeral ligaments), passing from the anterior aspect of the glenoid fossa to the neck of the humerus, and superiorly (coraco-humeral ligament) passing from the base of the coracoid process on to the tuberosities of the humerus. Both anteriorly and posteriorly the capsule gains added strength from the attachments of the **'rotator cuff' muscles** (subscapularis, supraspinatus, infraspinatus, teres minor), which are closely associated with the joint capsule. The weakest and least protected aspect of the capsule lies inferiorly.

Extrinsic ligaments are not thickenings of the capsule (**6.3.1**). The **coraco-acromial ligament** is triangular in shape. Its base is attached to the coracoid process while its apex passes to the tip of the acromion. It arches over the superior aspect of the joint, preventing upward dislocation of the humerus. Beneath this ligament is a large subacromial bursa which does not open into the shoulder joint. The **tendon of the long head of the biceps** (**6.3.2**) is attached to the supraglenoid tubercle and passes through the joint before giving rise to its muscular belly; this tendon helps to stabilize the head of the humerus within the joint during movement.

Qu. 3D *In which direction does a dislocation of the head of the humerus usually occur?*

Qu. 3E *Which nerve is most likely to be damaged by a dislocation of the shoulder joint? (see Seminar 12)*

It is a general principle that any artery lying close to a joint will supply it and its surrounding musculature. Nerves supplying muscles acting over a joint also supply proprioceptive and pain–sensing fibres to that joint (see p. 8); sympathetic vasomotor fibres supply arterioles of the synovial membrane.

Muscles acting on the shoulder joint

Examine the shoulder muscles which arise from the scapula. Those originating from its fossae are attached to only the medial two-thirds of the fossa (see p. 9):

Subscapularis (**6.3.3**) arises from the subscapular fossa and inserts into the lesser tubercle of the humerus; it also inserts into and thereby reinforces the capsule anteriorly. It forms most of the posterior wall of the axilla.

Qu. 3F *What are the actions of subscapularis?*

Qu. 3G *Which muscles other than subscapularis form the posterior wall of the axilla?*

Teres major arises from the dorsum of the scapula near the lower third of the lateral border, and passes laterally in front of the neck of the humerus to insert into the medial lip of the intertubercular groove of the humerus just below subscapularis (**6.3.3, 6.3.4**).

Supraspinatus (**6.3.4**) arises from the supraspinous fossa of the scapula and runs directly above the shoulder joint to insert into the capsule and the top of the greater tubercle of the humerus.

Qu. 3H *What is the major action of supraspinatus?*

Qu. 3I *Which other muscles act in combination with it?*

6.3.2
Interior of shoulder joint.

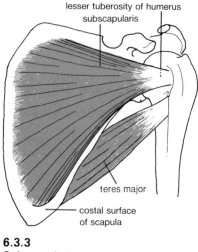

6.3.3
Subscapularis, teres major.

Infraspinatus (6.3.4) arises from the infraspinous fossa of the scapula and is attached to the capsule and the greater tubercle of the humerus between the supraspinatus and the teres minor. **Teres minor (6.3.4)** arises from the dorsal aspect of the lateral border of the scapula and is inserted into the greater tubercle of the humerus below infraspinatus.

Qu. 3J *What are the combined actions of infraspinatus and teres minor?*

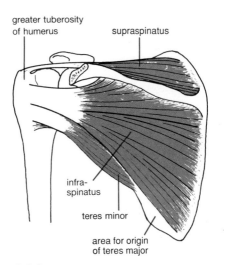

greater tuberosity of humerus

supraspinatus

infra-spinatus

teres minor

area for origin of teres major

6.3.4
Supraspinatus, infraspinatus, teres minor; acromion removed.

Deltoid (6.3.5) resembles an epaulette; it arises from the outer aspect of the lateral third of the clavicle, the acromion process, and the spine of the scapula, and inserts into the deltoid tuberosity halfway down the lateral aspect of the shaft of the humerus.

Qu. 3K *What is the action of this muscle?*

The attachments and actions of each muscle of the shoulder have been considered separately, but remember that no muscle acts alone. There are prime movers, antagonists, synergists, and fixators (see p. 9).

Particular attention should be paid in this prosection to the way in which the 'rotator cuff' muscles are blended together and their deeper parts fuse with the capsule of the shoulder joint to form the tendinous 'rotator cuff'. Because of the great mobility of the shoulder joint its capsule is necessarily lax, thus causing a potential instability of the joint. However, owing to the reinforcement of the capsule by the rotator cuff muscles which contract actively in movements of the shoulder, the joint is relatively stable in any position. When the head of the humerus dislocates it does so most frequently as a result of a downward blow on an abducted humerus. The head stretches the weakest (inferior) part of the capsule and comes to lie below either the coracoid process (anteriorly) or, less frequently, the spine of the scapula (posteriorly).

6.3.6 & 6.3.7 are a photograph and radiograph, respectively, of a young man who was injured playing rugby football. His left shoulder is very painful and he cannot move it. Instead of the normal smooth curve from acromion to deltoid there is a rather sharp angle. The deltoid lacks its normal bulging appearance. It looks 'empty', with one or two shallow dimples. Also, there is a bulge where the normal shoulder shows a depression, below the point at which the coracoid process should be palpable. Instead of the coracoid process, a large hard lump is to be felt at this site.

Qu. 3L *What injury has been sustained?*

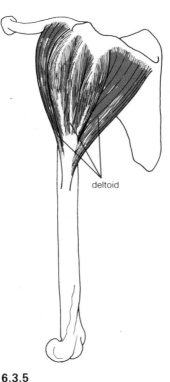

deltoid

6.3.5
Deltoid.

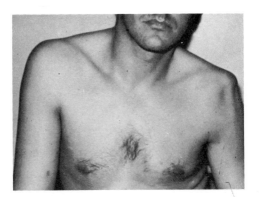

6.3.6
See Qu. 3L.

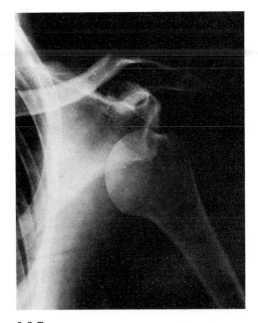

6.3.7
Radiograph of dislocation of left shoulder joint.

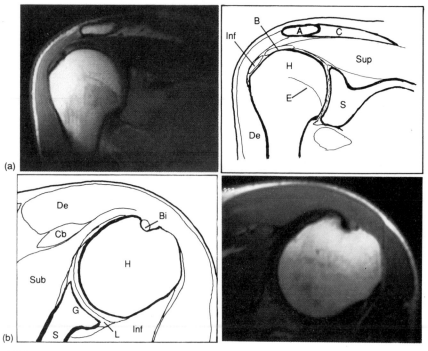

6.3.8
MRI of shoulder joint in coronal plane (a) and
horizontal plane (b). H—head of humerus;
E—epiphysial line; G—glenoid fossa;
L—posterior glenoid labrum; S—scapula;
A—acromion; C—clavicle; B—bursa;
De—deltoid; Cb—coracobrachialis;
Sup—supraspinatus; Inf—infraspinatus;
Sub—subscapularis; Bi—tendon of long head
of biceps in bicipital groove.

Now that strenuous exercise is considered beneficial, shoulder pain is a common complaint. Apart from fractures and joint disorders such as arthritis, the cause is often soft tissue damage which cannot be visualized by conventional radiology. Magnetic resonance imaging (MRI) is particularly useful in the diagnosis of such soft tissue damage, which includes tears of 'rotator cuff' muscle insertions, rupture of supraspinatus tendon, fatty degeneration of wasting muscles, and trapping of tissue between the head of the humerus and the coraco-acromial ligament. It is therefore important to be able to recognize normal MR images of the shoulder (**6.3.8**).

Qu. 3M *What soft tissues lie between the head of the humerus and the coraco-acromial ligament?*

After completing Seminars 9–12 make a note of the nerves supplying the muscles you have been studying.

Requirements:

Articulated skeleton and separate bones of the shoulder
Prosections of the shoulder joint with capsule intact and capsule opened
Prosections of the muscles of the shoulder
Synovial fluid (obtainable from butcher's shop or slaughterhouse)
Radiographs showing the shoulder joint in various positions.

Seminar 4

Elbow joint

Aims To study the movements which occur at the elbow joint; to consider the structural features of this 'hinge' joint that determine its movement and stability; and to study the muscles responsible for its movement.

A. Living anatomy

Note that the movements of your elbow joint are limited to <u>flexion and extension</u>. The joint can be flexed until the soft tissues of the arm and forearm contact one another, but cannot be extended beyond 180°. Any muscle which is attached to bones of the shoulder girdle or humerus, and which crosses the elbow joint anteriorly to insert into bones of the forearm, must necessarily flex the elbow. Conversely, those muscles which cross the posterior aspect of the elbow must extend the joint. Examine muscles of the anterior and posterior compartments of the arm by flexing and extending the elbow against resistance (see **6.2.6**). Note the position of the lateral and medial epicondyles and the olecranon process when the elbow is flexed; they form the corners of a roughly equilateral triangle (**6.4.1**). Malalignment of these landmarks suggests dislocation of the joint.

Place your arm naturally as if you were about to write or to perform some manual task; your elbow is probably semiflexed, and the forearm in the mid-prone position (p. 52). This is called the 'position of function'.

Sit down and get your partner to rest his semi-flexed forearm on your knee. Now hold your partner's elbow, and press with your left thumb over the tendon of biceps, then tap your thumb with a patellar hammer. This will stretch biceps and should cause its reflex contraction—a 'biceps jerk'. The extent of the reflex depends on the excitation of spinal cord motor neurons at C5 and C6, and on the integrity of the nerve which supplies biceps, its neuromuscular junction, and the muscle itself. Try also to elicit a 'triceps jerk' by tapping the triceps tendon just above the elbow. Compare the responses on the two sides of the body.

B. Radiology

Examine the radiographs of the elbow joint. Note the position of the bones in the extended (**6.4.2**) and flexed (**6.4.3**) position of the joint, and revise the secondary centres of ossification of the bones (pp. 36, 37).

In **6.4.3** draw a line along the middle of the shaft of the radius through the centre of the head of the

bone. Note that it passes through the centre of a circle which is outlined in part by the capitulum. In certain injuries, for instance a fracture of the shaft of the ulna, the head of the radius may be displaced from its position against the capitulum (**6.4.4**).

Draw a second line down the anterior border of the shaft of the humerus (**6.4.3**). This too should pass through the centre of the circle referred to above. The lower end of the humerus is normally angled 45° forward with respect to the shaft of the bone. If the lower end is displaced it is probably due to a supracondylar fracture of the shaft.

Extend your elbow and note the angle between the axis of the humerus and that of the forearm. This 'carrying angle' is greater in women than in men (see p. 35).

Qu. 4A *Examine* **6.4.5***: What abnormality can you see?*

C. Prosections

The elbow joint (6.4.6–6.4.11)

Examine a prosected elbow joint and note the shapes of the articular surfaces which are covered by hyaline articular cartilage; they consist of the quadrilateral-shaped **trochlea** of the humerus which articulates with the deep **trochlear notch** of the ulna, and the rounded **capitulum** at the lower end of the humerus, which articulates with the rounded concave upper aspect of the **head of the radius**. The articulation between the upper end of the radius and ulna is termed the **superior radio-ulnar joint** and will be considered again later. Note that the head of the radius also articulates on its medial aspect with the **radial notch** on the upper lateral aspect of the ulna. The head of the radius is held against the ulna by a sling-like **annular ligament** which is attached to either side of the radial notch. This ligament enables the head of the radius to rotate when its lower end is rotating around the lower end of the ulna in pronation and supination of the forearm.

Examine the **capsule** which is common to both the elbow and superior radio-ulnar joints. It is attached to the humerus immediately above the coronoid and radial fossae and to the inferior aspects of the medial and lateral epicondyles, thus avoiding the common origins of the superficial long flexor and extensor muscles of the forearm. Posteriorly it is attached to the *upper* margin of the olecranon fossa so that, when the elbow is straightened, the olecranon process of the ulna fits snugly into the fossa. Inferiorly, the capsule is attached to the annular ligament of the head of the radius so

6.4.1
Triangle formed by medial and lateral epicondyles and olecranon at elbow.

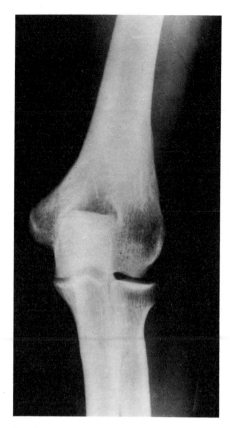

6.4.2
Left elbow joint, extended (AP).

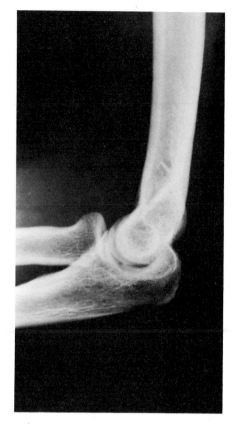

6.4.3
Left elbow joint, semiflexed (lateral).

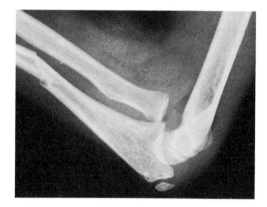

6.4.4
Fracture of shaft of ulna with displacement of head of radius in a child aged 14 years; note the secondary centre of ossification in the olecranon process.

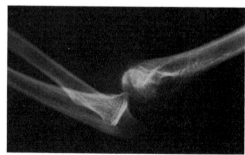

6.4.5
See Qu. 4A.

that movements of the upper end of the radius are not impeded; on the ulna, the capsule is attached to the rim of the trochlear notch.

The **synovial membrane** lines the capsule and all those parts of the joint which are not covered with hyaline articular cartilage; it also projects beneath the annular ligament of the head of the radius to surround the neck of the bone. As you might expect, the capsule of a hinge joint is reinforced medially and laterally by intrinsic **collateral ligaments**. The **lateral ligament** is fan-shaped and is attached to the lateral epicondyle superiorly and to the annular ligament inferiorly. The **medial ligament** is triangular in shape with three thickenings which are attached, in order, to the medial epicondyle superiorly, the coronoid process of the ulna inferiorly, and the medial side of the olecranon process.

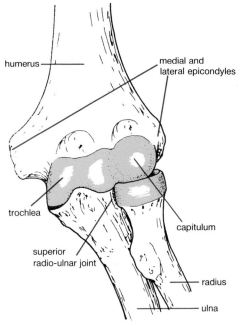

6.4.6
Bones of elbow joint (anterior aspect).

humerus

medial and lateral epicondyles

trochlea

capitulum

superior radio-ulnar joint

radius

ulna

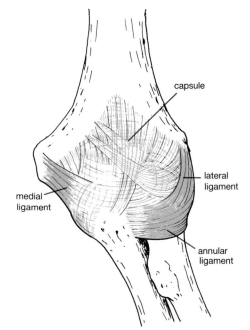

6.4.7
Capsule and ligaments of elbow joint (anterior aspect).

capsule

lateral ligament

medial ligament

annular ligament

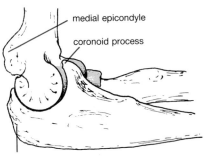

6.4.8
Bones of elbow joint (medial aspect).

medial epicondyle

coronoid process

olecranon process

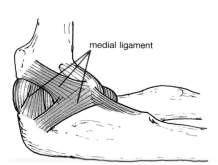

6.4.9
Capsule and ligaments of elbow joint (medial aspect).

medial ligament

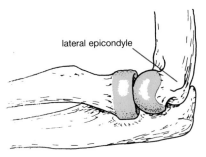

6.4.10
Bones of elbow joint (lateral aspect).

lateral epicondyle

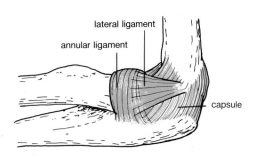

6.4.11
Capsule and ligaments of elbow joint (lateral aspect).

lateral ligament

annular ligament

capsule

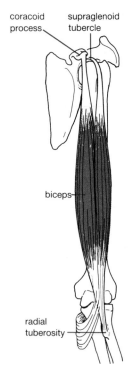

6.4.12
Biceps.

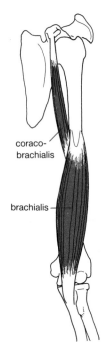

6.4.13
Coracobrachialis; brachialis.

Muscles of the anterior compartment of the arm

In the upper limb sheets of fascia (intermuscular septa) extend from the deep fascia, between the major groups of muscles, to the lateral and medial sides of the shaft of the humerus to which they are attached. In the forearm similar sheets of fascia pass from the deep fascia to the lateral and medial sides of the shafts of the radius and ulna respectively. This arrangement separates an anterior compartment group of muscles from a posterior compartment group of muscles. It reflects the basic embryological division of limb musculature into flexor (anterior) and extensor (posterior) groups. It also enables the muscles to obtain a wider origin from fascia as well as from bone.

Examine the muscles which lie in the anterior compartment of the arm:

Biceps (6.4.12) arises from the scapula by two heads, a **short head** from the tip of the coracoid process, and a **long head** from the supraglenoid tubercle within the capsule of the shoulder joint. The two heads join to form a large muscle belly which is inserted, by a flattened tendon passing in front of the elbow joint, into the posterior part of the tuberosity of the radius. An expansion of this tendon, the **bicipital aponeurosis**, crosses medially over the forearm superficial flexor muscles to be attached to the posterior border of the ulna via the deep fascia. In this way biceps exerts its flexor action on both bones of the forearm. In the 'position of function' biceps is a strong supinator and flexor of the forearm; it contracts strongly when a screw is being driven home or when a cork is being removed from a bottle. Since the short head crosses anterior to the shoulder joint it also has a flexor action at this joint, while the long tendon helps to stabilize the head of the humerus during movements of the shoulder joint.

Brachialis (6.4.13) arises from the lower half of the front of the humerus and is inserted into the front of the coronoid process of the ulna.

Coracobrachialis (6.4.13) is found in the upper arm and arises, with the short head of biceps, from the tip of the coracoid process; it is inserted into the medial aspect of the humerus half way down the shaft.

Qu. 4B *What are the actions of brachialis and coracobrachialis?*

Flex your pronated forearm against resistance and then repeat the movement with the forearm in a supinated position.

Qu. 4C *In which of these movements does biceps play a part?*

Muscles of the posterior compartment of the arm

The massive **triceps** muscle (**6.4.14**), as its name implies, arises from three separate heads, a **long head** from the infraglenoid tubercle of the scapula which lies outside the capsule of the shoulder joint, a **lateral head** from the posterior aspect of the shaft of the humerus above the radial groove, and a **medial head** which lies deeply and takes origin from the broad surface of the humerus below the groove. The three heads of the muscle unite to be inserted into the upper surface of the olecranon process of the ulna. **Anconeus**, really a part of triceps, passes obliquely from the lateral epicondyle to the lateral side of the olecranon process (see **6.6.7**). It acts to produce the slight abduction of the lower end of the ulna that occurs during pronation.

Qu. 4D *What are the actions of triceps?*

With the elbow flexed, extend the forearm against resistance as in **6.4.15**.

Qu. 4E *Which group of muscles contracts most strongly during extension against resistance?*

Next, with the elbow semiflexed and supported, and the hand holding a heavy weight, gradually extend the elbow joint.

Qu. 4F *Which group of muscles now contracts most strongly during extension?*

Finally, examine again the long head of biceps. Its tendon runs a prolonged course from the supraglenoid tubercle of the scapula over the head of the humerus within the shoulder joint to emerge beneath the capsule in the intertubercular groove. It then expands into its muscle belly, which often remains separate from the short head, until near the elbow. Within the shoulder joint the tendon is enclosed within a synovial sheath, which can become inflamed, causing pain. The tendon is also prone to spontaneous rupture due to degeneration of collagen within the tendon. Following a strain to biceps, which is not necessarily severe, the patient experiences a sudden pain in the front of the upper arm and a few days later some bruising appears.

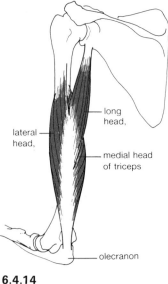

6.4.14
Triceps.

The contour of the biceps is altered. An example of this injury is seen in **6.4.16** but there is surprisingly little dysfunction since the short head is quite strong. There may be some weakness of supination of the forearm, but flexion of the elbow can still be performed by the short head and by brachialis.

Qu. 4G *With your knowledge of the position of the bony prominences around the elbow, how could you distinguish between a supracondylar fracture of the humerus and a dislocation of the ulna—without recourse to radiography?*

When you have completed Seminars 9–12 make a note of the nerve supply to the muscles you have been studying in this seminar.

Requirements:

Articulated skeleton and separate bones of the upper limb

Prosections of the elbow joint, muscles of the anterior and posterior compartments of the upper arm

Radiographs of the elbow joint of adults and children

Reflex-testing 'patellar' hammer.

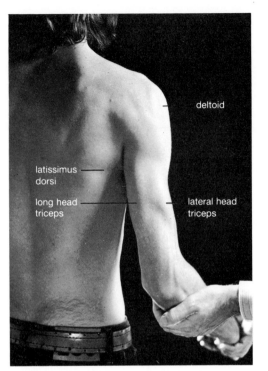

6.4.15
Triceps; demonstrated by extension of elbow against resistance.

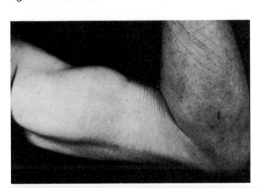

6.4.16
Ruptured long head of biceps.

Seminar 5

Joints of forearm and wrist

Aims To study movements at the radio-ulnar joints by which the hand can be 'rotated' on the arm; and the movements at the wrist which increase the range of movement of the hand; and to study the structural adaptations of all these joints for movement and stability.

A. Living anatomy

Movements of the forearm (pronation and supination), which enable the hand to be 'rotated' with respect to the humerus, occur at the **proximal** and **distal radio-ulnar joints**.

Keep your elbows to your sides and flex them to a right angle. Now rotate your forearms so that your palms face the floor—the forearms are now in the position of pronation. Next rotate your forearms so that your palms face the ceiling—they are now supinated. The range of movement is about 150°.

Qu. 5A *Against resistance, is pronation or supination the stronger? Why do screws have right-hand threads?*

Extend your arm and touch a wall in front of you with one extended finger; now pronate and supinate your forearm and do the same with each finger in turn. You will see that the forearm can rotate about an axis that can pass through any finger.

Qu. 5B *How is this possible?*

Now pronate and supinate the forearms freely without contact with the wall.

Qu. 5C *Through which finger does the axis of movement pass in this free movement?*

Hold your partner's hand as in a hand-shake. Supinate your forearm against your partner's resistance and feel the contraction of your biceps. Repeat this action while palpating over the upper end of the radius; you should be able to feel the contraction of the supinator muscle (see **6.6.13**). Now pronate your forearm against resistance and feel the contraction of a muscle (pronator teres) which arises from the medial epicondyle of the humerus. Rest your right forearm on a flat surface and supinate it. Mark the position of the styloid process of the ulna with your left index finger and then pronate your forearm. Note that the lower end of the ulna is abducted during pronation so that the axis of pronation/supination is through the centre of the wrist and palm.

When the forearm has to be immobilized, as after a fracture, the arm is usually held with the elbow semiflexed and the forearm mid-prone—the 'position of function'.

Movements of the wrist. With your right forearm in the supine position and with the elbow flexed, hold the distal ends of the radius and ulna with your left hand so that they cannot move.

Qu. 5D *What movements can you perform at the wrist joint?, and what is their extent?*

Note that all movements at the radio-carpal joint are accompanied by movements of the intercarpal joints. You have been testing the combined movements.

B. Radiology

Revise the radiographs of the forearm and hand which you studied in Seminar 1, noting in particular the articulations. **6.5.1 & 6.5.2** show the forearm in pronation and supination, respectively.

On radiographs **6.5.3** and **6.5.4** note the angle of slope of the distal articular surface of the radius in relation to its long axis in both antero-posterior and lateral views. In the antero-posterior view, if a line is drawn from the tip of the styloid process of the radius to the tip of the styloid process of the ulna, it lies at an angle of about 75° to the long axis of the shaft of the bones, the articular surface facing ulnarward. In a lateral view, a line drawn to join the margins of the carpal articular surface of the radius commonly slopes about 15° to the long axis of the shaft of the bone, the articular surface facing anteriorly.

The articular surface of the radius can be displaced after fractures of the radius resulting from falls on the wrist; it is important that the angulation is corrected if full movement of the wrist is to be restored.

Qu 5E *Examine radiographs 6.5.5 and 6.5.6. Which carpal bones articulate with the radius when the hand is adducted? and with the intra-articular disc when the hand is abducted?*

C. Prosections

Proximal and distal radio-ulnar joints

Both the proximal and distal radio-ulnar joints are synovial pivot joints. Examine the articular surfaces of these joints on a skeleton.

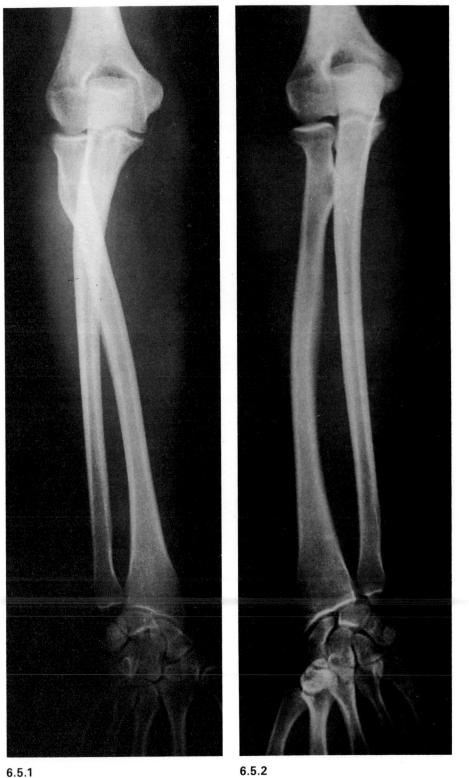

6.5.1
Forearm pronated (AP).

6.5.2
Forearm supinated (AP).

Examine a prosection of the **proximal radio-ulnar joint (6.5.7)** between the head of the radius and the radial notch on the proximal end of the ulna. The joint cavity and capsule are continuous with that of the elbow joint. The head of the radius is retained within a cup-shaped **annular ligament** attached to the two edges of the radial notch on the ulna. During pronation and supination of the forearm the head of the radius pivots on the capitulum of the humerus. If a child's forearm is

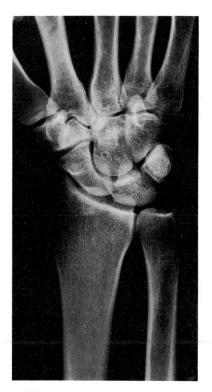

6.5.3
Wrist joint (AP).

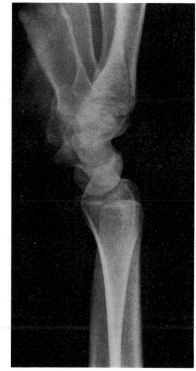

6.5.4
Wrist joint (lateral).

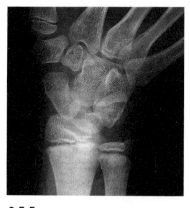

6.5.5
Wrist joint with hand adducted.

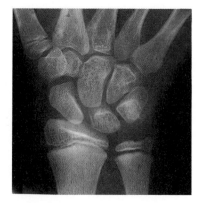

6.5.6
Wrist joint with hand abducted.

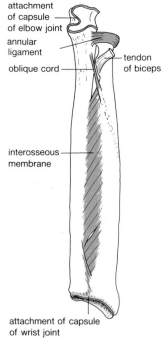

attachment
of capsule
of elbow joint

annular
ligament

oblique cord

tendon
of biceps

interosseous
membrane

attachment of capsule
of wrist joint

6.5.7
Capsule and ligaments of
radio-ulnar joints.

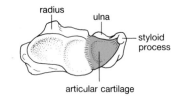

radius

ulna

styloid
process

articular cartilage

6.5.8
Intra-articular cartilage of inferior
radio-ulnar/wrist joint.

suddenly tugged (e.g. if it falls while holding its mother's hand) the head of the radius may be dislocated distally through the annular ligament.

Note the flexible sheet of collagenous tissue, the **interosseous membrane**, which unites the shafts of the radius and ulna. Its fibres pass distally from radius to ulna and thereby help to transfer compressive forces transmitted from the hand and radius to the ulna (and thus to the humerus). This effectively forms a mobile fibrous middle radio-ulnar joint.

Examine a prosection of the **distal radio-ulnar joint** (**6.5.7**) between the distal ends of the radius and ulna. Its capsule is continuous with that of the wrist joint, but its synovial cavity is usually separate. Examine carefully the triangular intra-articular disc (**6.5.8**) which is attached at its apex to the styloid process of the ulna, and at its base to the ulnar notch on the distal end of the ulna. This separates the distal end of the ulna from the cavity of the wrist joint.

Fractures of the distal ends of the radius and ulna are very liable to produce distortion of the distal radio-ulnar joint. Such distortion is likely to limit movement of the distal radio-ulnar joint, so that full pronation and supination movements may not

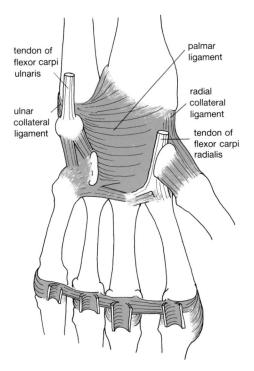

6.5.9
Capsule and ligaments of wrist, carpal, and metacarpo-phalangeal joints.

be possible. Loss of pronation can interfere with writing, and loss of supination with turning the hand over to receive something, or to open a door.

Wrist joint (6.5.9)

The wrist (radio-carpal) joint is a synovial ellipsoid joint. Examine its capsule—proximally it is attached around the articular margins of the distal ends of the radius and ulna while distally it is attached to the proximal row of carpal bones, i.e. the scaphoid, lunate, and triquetral. The attachments of the triangular intra-articular disc prevent the distal end of the ulna from articulating with the carpus, hence the wrist joint is more specifically termed the **radio-carpal joint**. Confirm the carpal bones which articulate with the radius and with the disc when the hand is adducted (**6.5.5**), and abducted (**6.5.6**). The capsule of the radio-carpal joint is strengthened by dorsal and palmar 'ligaments'; collateral ligaments are attached, on the ulnar side, from the styloid process of the ulna to the pisiform and triquetral bones and, on the radial side, from the styloid process of the radius to the scaphoid.

A Colles' fracture, named after the surgeon who described the injury in the years before X-rays were discovered, is the most common fracture occurring at the wrist (**6.5.10**, **6.5.11**). During a fall on the outstretched hand, especially in the aged in whom the bones are relatively less resistant, the strength of the palmar part of the capsule causes the upward, backward, and lateral displacement of the distal part of the radius which is typical of a Colles' fracture.

Requirements:

Articulated skeleton of upper limb
Prosections of proximal and distal radio-ulnar joints, and wrist joint
Radiographs of these joints.

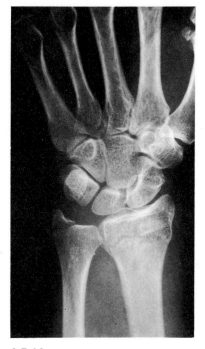

6.5.10
Fractured lower end of radius and ulna (Colles' fracture; AP).

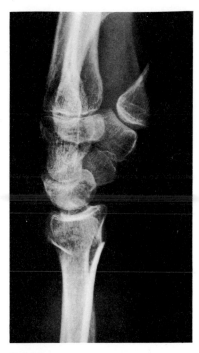

6.5.11
Fractured lower end of radius and ulna (Colles' fracture; lateral).

Seminar 6

Muscles and movements of forearm, wrist, and hand

Aims To study the groups of muscles in the forearm which are responsible for the wide range of coarse and fine movements of which the distal extremity of the upper limb is capable.

When describing the forearm it is often helpful to use the terms 'ulnar' and 'radial' rather than medial and lateral, because pronation can obscure the basic anatomical position.

A. Living anatomy

The actions of the forearm muscles are conveniently investigated as you identify each muscle in the dissected material.

B. Prosections

Muscles of the anterior compartment of the forearm
(6.6.1, 6.6.2)

The anterior compartment of the forearm contains flexor muscles of the wrist and fingers, and muscles which produce pronation and supination of the forearm. Some muscles also have an action on the elbow joint by virtue of their attachment to the humerus.

At the front of the elbow, the muscles define a triangular area referred to as the **cubital fossa**, bounded medially by the lateral border of pronator teres, laterally by brachioradialis, and above by an imaginary line between the medial and lateral epicondyles (**6.6.1** and see **6.8.2**).

At the wrist the tendons of most muscles which insert in the hand pass under the fibrous **flexor retinaculum** enclosed in synovial sheaths (**6.6.14**).

Examine first the **common origin of the superficial flexor muscles of the forearm** which is attached to the medial epicondyle.

Identify **pronator teres** (**6.6.3**) which arises by two heads from a) the common flexor origin and the lower end of the medial supracondylar ridge and b) the medial border of the coronoid process of the ulna, and passes diagonally across the forearm to insert half-way down the front of the shaft of the radius. As its name suggests, it pulls the radius across the ulna thus pronating the forearm. When in Seminar 10 you study the median nerve you will

see that it passes between the two heads of pronator teres as it leaves the cubital fossa to enter the forearm.

Flexor carpi radialis arises from the common flexor origin and crosses the wrist to insert into the bases of the 2nd and 3rd metacarpal bones.

Palmaris longus, which is not always present, arises from the common flexor origin and is inserted into the very thick, tough fascia—the palmar aponeurosis—which lies beneath the skin of the palm.

Flexor carpi ulnaris arises not only from the common flexor origin but also from the medial side of the olecranon and the posterior subcutaneous border of the ulna; it inserts into the pisiform bone with an extension on to the base of the 5th metacarpal.

Flexor digitorum superficialis (**6.6.4**) lies on a deeper plane than the other superficial flexors. It arises from the common flexor origin, from the coronoid process of the ulna, and from the front of the upper part of the shaft of the radius by a thin aponeurosis. Near the wrist the muscle divides to give four tendons which pass beneath the flexor retinaculum. In the fingers these tendons divide to allow the tendons of flexor digitorum profundus (see below) to pass through to reach the distal phalanges; the divided tendon slips of flexor digitorum superficialis partially reunite and insert into the base and sides of the middle phalanges.

Flex your wrist against resistance and identify the tendons on the anterior aspect of your forearm (**6.6.5**).

Qu. 6A *What are the (prime mover) actions of these individual muscles?*

Now examine the deep muscles of the forearm.

Flexor digitorum profundus (**6.6.6**) lies deeply in the forearm. It originates from the medial and anterior aspects of the upper shaft of the ulna and the adjoining interosseous membrane; its tendons cross the wrist and the palm, and into the fingers where they pass through the divided tendons of flexor digitorum superficialis to insert into the base of the distal phalanges.

Qu. 6B *What simple test could you devise to investigate the functional integrity of a) flexor digitorum superficialis, b) flexor digitorum profundus?*

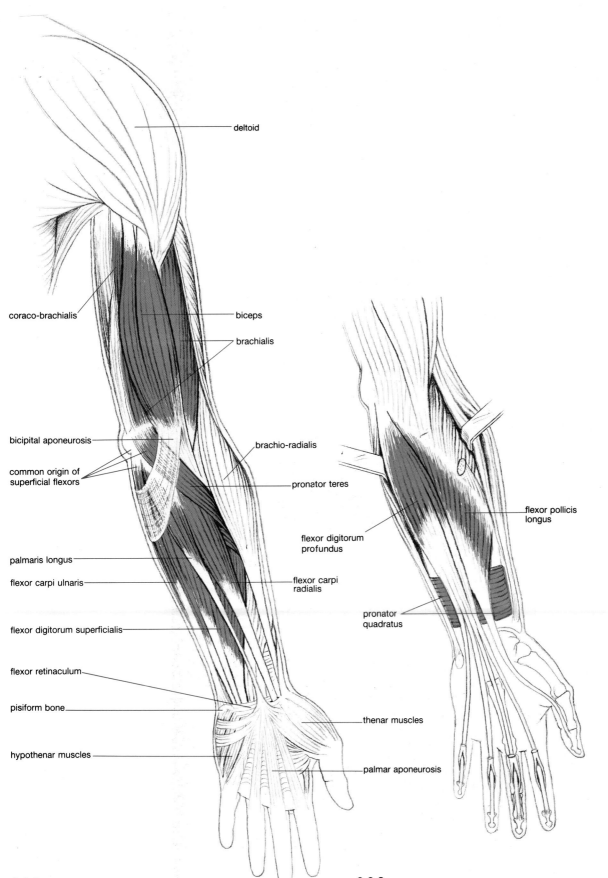

deltoid

coraco-brachialis

biceps

brachialis

bicipital aponeurosis

common origin of
superficial flexors

brachio-radialis

pronator teres

palmaris longus

flexor carpi ulnaris

flexor digitorum superficialis

flexor retinaculum

pisiform bone

hypothenar muscles

flexor carpi
radialis

flexor pollicis
longus

flexor digitorum
profundus

pronator
quadratus

thenar muscles

palmar aponeurosis

6.6.1
Superficial flexor muscles of arm and forearm.

6.6.2
Deep flexor muscles of forearm.

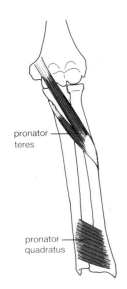

pronator teres

pronator quadratus

6.6.3
Pronator muscles.

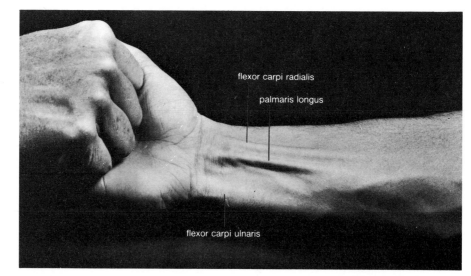

flexor carpi radialis

palmaris longus

flexor carpi ulnaris

6.6.5
Long flexor tendons; demonstrated by flexion of wrist against resistance.

Flexor pollicis longus (6.6.6) arises from the anterior surface of the radius between the attachments of pronator teres and pronator quadratus. This strong muscle forms a tendon which receives fleshy fibres down to the wrist, and passes beneath the flexor retinaculum to insert into the base of the distal phalanx of the thumb.

In each digit, the tendons of the long flexor muscles lie in a synovial sheath encased by a **fibrous flexor sheath** attached to the sides of the phalanges. The fibrous flexor sheaths are strong over the shafts of phalanges, but flexible over the joints; they allow the long flexor tendons to exert their pull without 'bowstringing' away from the phalanges.

Pronator quadratus (6.6.3) is attached anteriorly to the distal ends of the shafts of the radius and ulna. With pronator teres it produces pronation of the forearm.

Qu. 6C *When the forearm is being flexed would you expect muscles involved with pronation and supination also to be active (i.e. acting synergistically)?*

Muscles of the posterior compartment of the forearm
(6.6.7, 6.6.8)

The posterior compartment of the forearm contains both the extensor muscles which act on the wrist and fingers, and the supinator. Many of the superficial extensors arise from a **common extensor origin** on the *anterior* aspect of the lateral epicondyle of the humerus. Those muscles which insert in the hand form tendons which pass beneath an **extensor retinaculum** at the wrist.

Brachioradialis (6.6.9) arises from the upper part of the lateral supracondylar ridge and passes down the radial border of the forearm, overlapping both flexor and extensor compartments, to insert on the lateral aspect of the distal end of the radius. It is a moderately strong flexor of the elbow, and tends to bring the forearm from either extreme

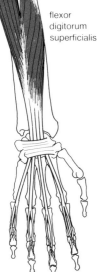

flexor digitorum superficialis

6.6.4
Flexor digitorum superficialis.

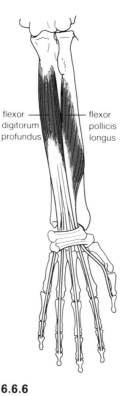

flexor digitorum profundus

flexor pollicis longus

6.6.6
Flexor digitorum profundus; flexor pollicis longus.

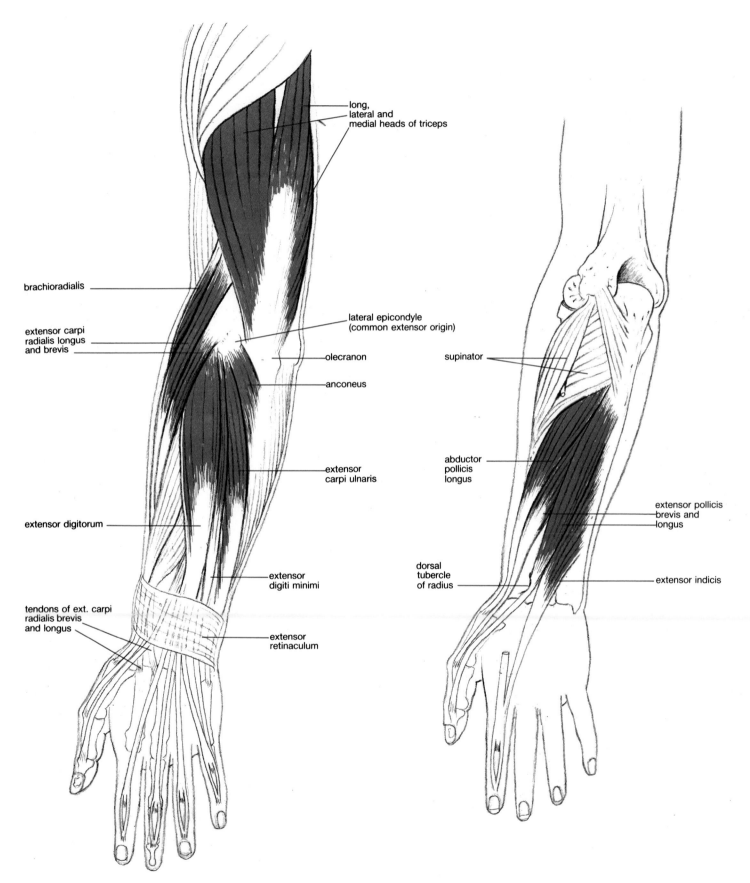

long,
lateral and
medial heads of triceps

brachioradialis

extensor carpi
radialis longus
and brevis

lateral epicondyle
(common extensor origin)

olecranon

anconeus

extensor
carpi ulnaris

extensor digitorum

extensor
digiti minimi

tendons of ext. carpi
radialis brevis
and longus

extensor
retinaculum

supinator

abductor
pollicis
longus

extensor pollicis
brevis and
longus

dorsal
tubercle
of radius

extensor indicis

6.6.7
Superficial extensor muscles of arm and forearm.

6.6.8
Deep extensors of forearm.

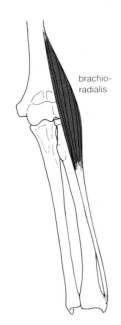

6.6.9
Brachioradialis.

6.6.10
'Mallet finger'.

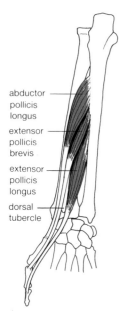

6.6.11
Long extensor group muscles of thumb.

supination or pronation to the mid-prone position ('position of function').

Extensor carpi radialis longus arises from the lower part of the lateral supracondylar ridge of the humerus and inserts into the dorsal aspect of the base of the 2nd metacarpal.

Extensor carpi radialis brevis arises from the common extensor origin and inserts into the dorsal aspect of the base of the third metacarpal.

Extensor carpi ulnaris arises from the common extensor origin and inserts into the dorsal aspect of the base of the 5th metacarpal.

These latter three muscles are all powerful extensors of the wrist and act synergistically with the long flexors of the fingers in the power grip (p. 63). Note the symmetrical insertion of the carpal flexors and extensors; when they contract together, the wrist is fixed so that delicate movements of the fingers can occur from a stable base.

Qu. 6D *Which muscles of the arm are acting when a glass is raised to the lips?*

Now examine **extensor digitorum** and **extensor digiti minimi** which also arise from the common extensor origin. The tendons of these muscles pass down the forearm and over the back of the wrist and hand into the fingers. On the back of the hand the tendons are usually interconnected. Place your hand palm downwards on a flat surface and extend each of your fingers in turn; you will then be able to locate the tendinous interconnections. Note from the prosection that the tendons are inserted into triangular **fibrous dorsal extensor expansions** which wrap around the dorsal and lateral aspects of the proximal phalanges (see **6.7.7**). These extensor expansions are each inserted by a central slip into the dorsal surface of the base of the middle phalanx, and by two conjoined lateral slips into the dorsal surface of the base of the terminal phalanx. Some small muscles of the palm of the hand (lumbricals and interossei; p. 66) gain attachment to the extensor expansions. If an extensor tendon is torn from its insertion to the distal phalanx, the condition of 'mallet finger' results (**6.6.10**). The terminal interphalangeal joint is flexed about 45° and, although it will flex both actively and passively, and extend passively, the patient cannot straighten the joint actively. The injury follows a passive flexion force on the distal interphalangeal joint when the extensor tendon is under strain.

Degenerative changes can arise in the fibrous tissue near the common extensor origin. Sharp local tenderness develops in the muscle mass just in front of the lateral epicondyle, and pain is caused when the wrist and fingers are actively extended against resistance. The lesion has been called 'tennis elbow' from its supposed causation by backhand strokes, but in fact very few sufferers seem to play this sport.

Examine the deep extensor muscles of the forearm which are attached largely to the ulna and the interosseous membrane and pass diagonally across

the back of the forearm to be inserted into the dorsum of the bones of the thumb and index finger (**6.6.8**).

Abductor pollicis longus (**6.6.11**) is inserted through its long tendon into the base of the 1st metacarpal while **extensor pollicis brevis** inserts into the base of the proximal phalanx of the thumb; their tendons lie together on the radial aspect of the wrist. The tendon of **extensor pollicis longus** hooks around the dorsal (Lister's) tubercle of the radius, to be inserted into the base of the terminal phalanx of the thumb. If you extend your thumb, the tendons of these muscles will be clearly seen (**6.6.12**). Between them, they form a hollow at the base of the thumb into which snuff can be placed and inhaled—hence the time-honoured name 'anatomical snuffbox'. In its depths you should be able to feel the pulsations of the radial artery which winds around the wrist before passing between the heads of the 1st dorsal interosseous muscle (p. 70) to enter the palm.

Spontaneous rupture of the tendon of extensor pollicis longus may occur and is thought to be due to ischaemia resulting from diseased interosseous vessels. The rupture occurs in the tunnel beneath the extensor retinaculum as the tendon changes direction around the dorsal tubercle of the radius. The patient feels that the thumb has 'dropped'; the distal phalanx of the thumb will not extend properly, and it may get in the way of the fingers sufficiently to require amputation.

Deep in the upper part of the forearm locate **supinator** (**6.6.13**). This muscle surrounds the upper third of the radius. Its superficial fibres arise from the lateral epicondyle of the humerus, its deep fibres from the lateral border of the upper part of the ulna (the supinator crest) and from adjacent ligaments. The ulnar fibres pass behind the radius, join the deep aspect of the humeral fibres, and insert into the proximal third of the anterior aspect of the radius, just proximal to pronator teres. The deep branch of the radial nerve (p. 85) passes between the two layers to gain access to the posterior compartment of the forearm.

When you have completed Seminars 9–12 make a note of the nerve supply to the muscles you have been studying.

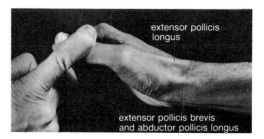

6.6.12
'Anatomical snuffbox.'

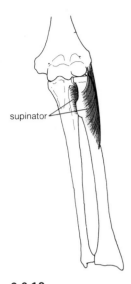

6.6.13
Supinator.

Control of the long tendons at the wrist

The **flexor** and **extensor retinacula** are thickened fibrous bands of deep fascia which hold the tendons firmly to the wrist and palm and prevent 'bow-stringing'.

The **flexor retinaculum** (**6.6.14**) is a thick band which converts the concavity of the palmar surface of the carpus into an osteo-fascial channel, the **carpal tunnel**. It is attached to the pisiform and hook of hamate medially, and to the tubercle of the scaphoid and ridge on the trapezium laterally. The long flexor tendons to the thumb and fingers together with the tendon of flexor carpi radialis and the median nerve pass through the carpal tunnel covered with synovial sheaths.

The **extensor retinaculum** (**6.6.7**) stretches across the back of the wrist and converts the grooves on the dorsum of the distal end of the radius into separate channels for the long extensor tendons and their synovial sheaths (see **6.7.13**). It is attached medially to the pisiform bone and hook of the hamate (as is the flexor retinaculum) and laterally to the radius.

Qu. 6E *Why is the extensor retinaculum not attached to the dorsal aspect of both the radius and the ulna?*

Requirements:

Articulated skeleton and separate bones of the upper limb
Radiographs of elbow, wrist, and hand
Prosections of anterior and posterior compartments of the forearm showing a) superficial and b) deep muscles
Prosections of the flexor and extensor retinacula, fibrous flexor sheaths, and dorsal digital expansions.

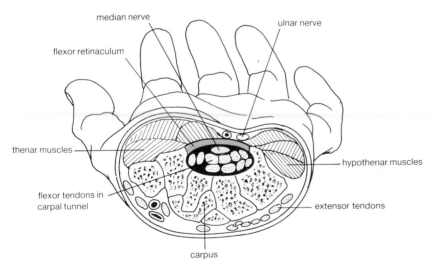

6.6.14
Carpal tunnel and contents (tendon sheaths not shown; transverse section).

Seminar 7

Joints, muscles, and movements of the hand

Aims To study the hand and the musculoskeletal elements which enable it to grip and manipulate objects of various shapes and sizes, and to gesticulate; to consider the hand as an actively exploratory sensory organ.

A. Living anatomy

Rest the back of your hand on a flat surface and note that the index, middle, ring, and little fingers are progressively more flexed. Note also that the axis of the thumb, and therefore also of its movements, lies at right angles to the other digits (i.e. its nail is directed laterally). This is the natural position of the hand.

Qu. 7A *What are the advantages of this position?*

A deep laceration which severs the long flexor tendons will distort the pattern and the injury can be diagnosed on sight.

Movements of the fingers and thumb (6.7.1)

The functional axis of the hand passes through the centre of the palm and middle finger. Abduction and adduction of the fingers are related to this plane. The middle finger can therefore only be abducted from the midline.

Qu. 7B *What movements can you make at the metacarpo-phalangeal joints? (6.7.1) and the interphalangeal joints? and what is their extent?*

Flex the thumb across the palm and then extend it; abduct it at right angles away from the index finger and then adduct it to its original position. Now approximate the pads of the thumb and little finger; this movement of **opposition** is, with respect to its range, unique to Man. It is achieved by a combination of movements which effectively rotate the thumb at its carpo-metacarpal joint.
  Surgical repair of the hand after accidents causing loss of the thumb entails fashioning a mobile stump to enable at least partial opposition to be restored.

Qu. 7C *Why is an ability to oppose the pads of the thumb and the fingers of such importance to Man?*

Grasp the head of each metacarpal in turn and assess its mobility. The metacarpal of the thumb (1st) is the most mobile, followed by the 5th and

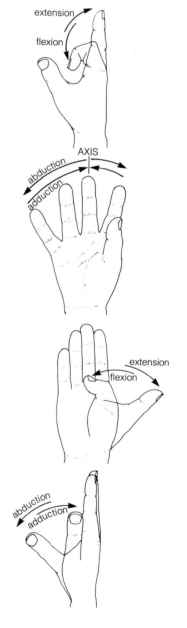

6.7.1
Movements of fingers and thumb.

4th; the second and third metacarpals are practically immobile. This mobility of the metacarpals allows the palm to be curved around an object it is holding.

Qu. 7D *During the evolutionary development of opposition and other manipulative movements of the hand and fingers, evolution of the neural control of these movements was taking place. What might this involve?*

Grips

First pick up a heavy tool such as a hammer and note the position of your hand and wrist. This is the **power grip (6.7.2)**: the handle of the tool is gripped firmly between the opposed thumb on one side and the four fingers on the other. The wrist is extended by about 45° and, if forcibly flexed, the grip becomes much weaker and can be broken.

Pick up a briefcase by its handle. Your hand will be making a **hook grip (6.7.3)** in which the handle of the case is supported by the flexed fingers. The thumb may be opposed, but this is not essential, and the wrist is not extended.

Pick up a pen as if you were about to write with it; this is the **precision grip (6.7.4)**. The pen is held between the opposed thumb and forefinger and supported against the middle finger; the ring and little fingers are flexed.

Finally, hold a key as if you were about to insert it into a lock. Your hand adopts the **key grip (6.7.5)** by which the object is gripped between the opposed thumb and the radial side of the flexed index finger.

There are, of course, many variations on these basic grips.

Revise the position and attachments of the **flexor retinaculum** (p. 61), and mark it on your palm with a skin pencil. Note that it lies distal to the distal skin crease of the wrist. Note too that the fleshy thenar and hypothenar eminences, which are formed by the small muscles of the thumb and little finger, overlap and originate from the retinaculum.

Remind yourself also of the position of the tendons of the long flexor and extensor muscles as they cross the wrist to their insertions into the bones of the carpus and the phalanges. Identify them on your own wrist (see **6.6.5**).

B. Radiology

Review the radiological appearance of the bones of the hand (p. 37; **6.1.10–6.1.13**) and be sure you can identify the individual carpal bones. Note again the overlapping articulation of the distal row of carpal bones and the bases of the metacarpals.

C. Prosections

Joints of the hand

Study a prosection of the joints of the hand.

Intercarpal joints. Adjacent carpal bones articulate together by plane synovial joints which enable small amounts of sliding movements to occur, but only between the proximal and distal row of carpal bones does a significant amount of movement occur. Movements at this **mid-carpal joint** complement those occurring at the wrist joint.

Carpo-metacarpal joints. Clench your fist and observe the movements of your knuckles, which reflect those occurring at synovial joints at the base of the 2nd–5th metacarpals. Only the 4th and 5th metacarpals move to any extent, and enhance the grip. The **carpo-metacarpal joint of the thumb (6.7.6)** is a synovial 'saddle' joint specialized to allow opposition of the thumb. The articular surfaces on the trapezium and on the base of the 1st metacarpal are both concave in one axis and convex in the other (like a rider in a saddle).

Metacarpo-phalangeal joints. If the metacarpo-phalangeal joints of the index, middle, ring, and little fingers are extended, a certain amount of side-to-side movement of the fingers can occur. This is described as abduction and adduction in relation to an axis through the middle finger (**6.7.1**). When the metacarpo-phalangeal joints are flexed, these movements are much restricted. This is because, in lateral profile, the surface of the head of a metacarpal is shaped like a cam (**6.7.7, 6.7.8**). The collateral ligaments of each joint pass from dorsal tubercles on the head of the metacarpal to lateral tubercles at the base of the proximal phalanx. If the metacarpo-phalangeal joints are extended these ligaments are lax and abduction and adduction movements are possible. In flexion, the collateral ligaments are tightened thereby preventing side-to-side motion at the joint. This increases the stability of grip during the power grip (**6.7.2**). If an injured hand is immobilized with the metacarpo-phalangeal joints fully extended, the collateral ligaments may shorten in their relaxed state. The patient may then find it impossible to make a satisfactory grip because the collateral ligaments become too taut, limiting full flexion. Therefore, if a hand has to be put in plaster due to injury, the metacarpo-phalangeal joints of the ulnar four digits are immobilized in nearly full flexion, so that shortening of the collateral ligaments cannot occur, and a normal range of movement is obtained when mobilization begins.

The heads of metacarpals 2–5 are held together by a **deep transverse metacarpal ligament** which is attached to the palmar aspect of the metacarpo-phalangeal joints.

Interphalangeal joints. There is little cam effect at these synovial hinge joints (**6.7.7**). Their collateral ligaments are taut throughout the range of flexion and extension.

In general, after injuries, the fingers are best immobilized so that the metacarpo-phalangeal joint is flexed to 80°, and the interphalangeal joints to 10°.

6.7.2
Power grip.

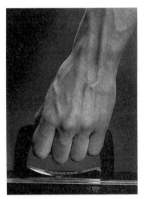

6.7.3
Hook grip.

6.7.4
Precision grip.

6.7.5
Key grip.

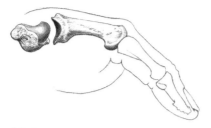

6.7.6
Carpo-metacarpal joint of thumb ('saddle joint').

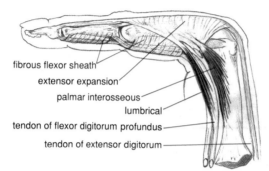

fibrous flexor sheath
extensor expansion
palmar interosseous
lumbrical
tendon of flexor digitorum profundus
tendon of extensor digitorum

6.7.7
Lateral view of right index finger showing metacarpo-phalangeal ('cam') joint and interphalangeal joints, flexor tendons, and fibrous flexor sheaths. Note the insertion of lumbrical and interosseous tendons into the dorsal extensor expansion.

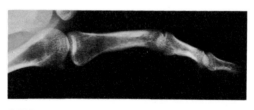

6.7.8
Metacarpo-phalangeal and interphalangeal joints of index finger (lateral).

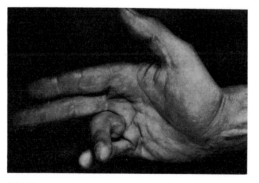

6.7.9
Dupuytren's contracture.

Muscles and tendons of the hand

Muscles and tendons in the hand are said to form four layers. The most superficial (first layer) are muscles of the thenar and hypothenar eminence; beneath them, and lying centrally in the palm, are the long flexor tendons and the lumbrical muscles that arise from them (second layer); adductor pollicis originates deep to the long flexor tendons and forms the third layer; and the palmar and dorsal interosseous muscles form the deepest (fourth) layer. All the muscle tissue is either on the palmar aspect of, or lies between the metacarpal bones; only tendons are found on the dorsum of the hand.

Examine first the form and distribution of the **palmar aponeurosis (6.6.1)**. Its thick central portion lies in the palm between the muscles of the thenar and hypothenar eminences, separated from the skin of the palm by a layer of fat divided into multiple little loculi by fibrous tissue. It merges with the flexor retinaculum and, if it is present, with the tendon of palmaris longus which crosses the retinaculum. The aponeurosis divides into four slips, one to each of the fingers. Each slip of the aponeurosis is attached to the fibrous sheath of the finger and to the deep transverse metacarpal ligament on either side. Other fibres pass along the sides of the proximal phalanges to the base of the middle phalanges. Its very thin lateral parts cover the thenar and hypothenar muscles.

The palmar aponeurosis is commonly affected by a condition called Dupuytren's contracture (**6.7.9**) in which a dense fibrosis occurs in the aponeurosis, particularly the fibres to the little and ring fingers. Thickening of the fibrous tissue occurs and a flexion contracture of the metacarpo-phalangeal and proximal interphalangeal joints follows. This may cause severe deformity requiring operative excision of the thickened, contracted aponeurosis.

At the roots of the fingers, a thin **superficial transverse metacarpal ligament** stretches across the palm in the superficial fascia, attached to the skin of the finger clefts. Vessels and nerves passing into the fingers pass between it and the deep transverse metacarpal ligament.

Examine the three superficial muscles which form the **thenar eminence (6.7.10, 6.7.11)**. **Abductor pollicis brevis** arises from the scaphoid and the flexor retinaculum and inserts into the lateral side of the base of the proximal phalanx of the thumb; as its name suggests, it abducts the thumb. **Flexor pollicis brevis** is more medially placed; it arises from the trapezium and the flexor retinaculum and inserts with the abductor pollicis brevis; it flexes the thumb across the palm. **Opponens pollicis** lies deep to the former two muscles; it arises from the trapezium and the flexor retinaculum and passes obliquely to insert into the whole length of the lateral border of the metacarpal of the thumb; it brings the pad of the thumb into contact (opposition) with the pad of any of the fingers. The three superficial muscles of the thenar eminence are

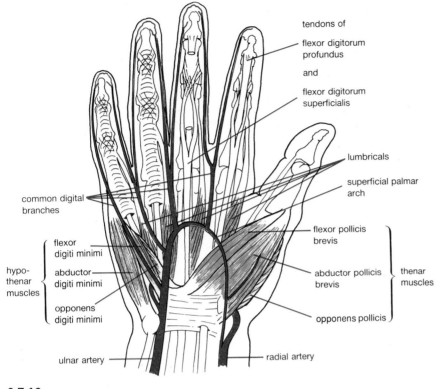

tendons of
flexor digitorum profundus
and
flexor digitorum superficialis

lumbricals

superficial palmar arch

common digital branches

flexor pollicis brevis

flexor digiti minimi

abductor digiti minimi

abductor pollicis brevis

hypothenar muscles

thenar muscles

opponens digiti minimi

opponens pollicis

ulnar artery

radial artery

6.7.10
Superficial aspect of palm.

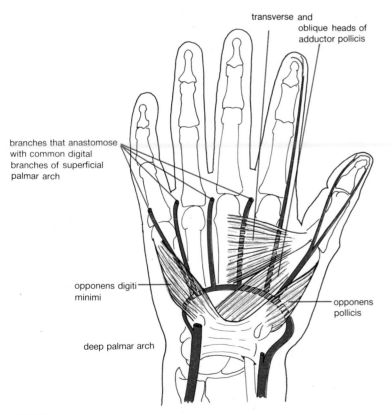

transverse and oblique heads of adductor pollicis

branches that anastomose with common digital branches of superficial palmar arch

opponens digiti minimi

opponens pollicis

deep palmar arch

6.7.11
Deep aspect of palm.

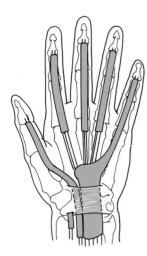

6.7.12
Synovial sheaths of flexor tendons.

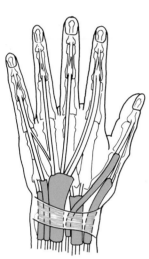

6.7.13
Synovial sheaths of extensor
tendons.

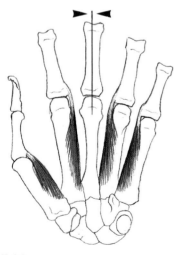

6.7.14
Palmar interossei.

supplied by the median nerve. Deep to them, and separated by the tendon of flexor pollicis longus, lies adductor pollicis (see below).

Examine next the three small muscles of the **hypothenar eminence (6.7.10, 6.7.11)** which correspond with those of the thenar eminence. **Abductor digiti minimi** arises from the pisiform bone and inserts into the medial side of the base of the proximal phalanx; it abducts the little finger. **Flexor digiti minimi** lies lateral to the abductor; it arises from the hook of the hamate and the flexor retinaculum and inserts with abductor digiti minimi; it flexes the metacarpo-phalangeal joint of the little finger. **Opponens digiti minimi** lies deep to the flexor and opponens muscles; it also takes origin from the hook of the hamate and the flexor retinaculum but is inserted into the whole length of the medial aspect of the shaft of the metacarpal bone; it opposes the pad of the little finger slightly toward that of the thumb.

Identify again the **tendons** of the long flexor muscles of the fingers and thumb in the forearm, and trace them through the carpal tunnel beneath the flexor retinaculum. Note that the carpal tunnel contains not only the long flexor tendons but also the median nerve (p. 80) passing into the hand.

Remove the palmar aponeurosis and trace the tendons of the long flexors as they pass across the palm toward the fingers. Find the four small worm-like **lumbrical** muscles **(6.7.7, 6.7.10)** that arise from the radial side of the tendons of flexor digitorum profundus and pass into the finger webs to insert into the base of the extensor expansions. The lumbricals are supplied by the same nerves that supply the part of flexor digitorum profundus from which they arise—the medial two by the ulnar nerve and the lateral two by the median nerve.

Qu. 7E *What is the action of the lumbrical muscles on (a) the metacarpo-phalangeal joint, (b) the interphalangeal joints?*

Beneath the flexor retinaculum the tendons of flexor carpi radialis and flexor pollicis longus have their own synovial sheaths, but the long flexor tendons to the fingers are surrounded by a **common flexor synovial sheath (6.7.13)**. This extends from about 3 cm proximal to the wrist to the level of the heads of the metacarpal bones, but is separate from the synovial sheaths that surround the tendons within the fibrous flexor sheaths of the fingers, except in the case of the little finger. The synovial sheaths are double-layered with a thin film of synovial fluid between the layers to reduce friction. Compare the different arrangement of the extensor synovial sheaths (**6.7.13**).

Qu. 7F *What might be the danger if a deep wound of the little finger became septic?*

Examine the **fibrous flexor sheaths (6.7.7, 6.7.10)** which hold the long flexor tendons in place. The 'sheaths' are really tunnels of fibrous tissue attached to the sides of the phalanges of the fingers and thumb. They are strong over the phalanges,

but much thinner over the joints so that they do not impede finger movement.

Movement of the long flexor tendons within the fibrous flexor sheaths is lubricated by digital **synovial sheaths** which surround the tendons. Fine blood vessels covered with synovial membrane (vincula) pass from the phalanges to the tendons before they insert.

Pull the long tendons to one side or cut through them to examine **adductor pollicis (6.7.11)**. It has two heads of origin, one from the shaft of the middle (3rd) metacarpal, the other from the bases of the 2nd and 3rd metacarpals and the adjacent carpal bones. The muscle is inserted into the medial (ulnar) side of the base of the proximal phalanx of the thumb. Its function is to bring the abducted thumb, which lies at right angles to the palm, back into contact with the index finger.

Deep to all of these locate the **interosseous** muscles which occupy the spaces between the shafts of the metacarpal bones and which abduct and adduct the metacarpo-phalangeal joints. The **palmar interossei (6.7.14)** are small unicipital muscles which arise from the ulnar side of the 1st and 2nd metacarpals and from the radial side of the 4th and 5th metacarpals; they pass around the side of the proximal phalanx to which they are related to insert into the base of the dorsal extensor expansions. They adduct the fingers towards the axis of the hand which runs through the middle finger; they also flex the metacarpo-phalangeal joints and extend the phalanges. The **dorsal interossei (6.7.15)** are larger and take origin by two heads, one from each of the metacarpal bones between which they lie. They insert into the dorsal digital expansion and into the base of the proximal phalanges of digits 2, 3 (both sides), and 4. They act to abduct these fingers away from the axis of the hand, but share the other actions of the palmar interossei.

Adductor pollicis and all the interossei are supplied by the deep branch of the ulnar nerve (p. 82) which runs on their palmar surface.

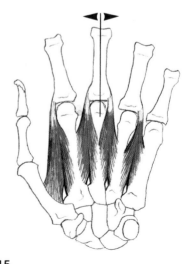

6.7.15
Dorsal interossei.

Fascial spaces of the palm

The palm is divided into a number of **palmar spaces** by fibrous tissue septa (**6.7.16**, **6.7.17**). From the medial border of the triangular palmar aponeurosis the **medial palmar septum** passes deeply between the hypothenar muscles and the long flexor tendons to the little finger to attach to the 5th metacarpal. Similarly, from the lateral border of the aponeurosis the **lateral palmar septum** passes between the three superficial thenar muscles and the tendon of flexor pollicis longus to attach to the 1st metacarpal. These two septa separate the thenar and hypothenar eminence compartments from the palmar spaces. Between the palmar aponeurosis and the metacarpal bones with their attached interosseous muscles, the central part of the palm contains the long flexor tendons and lumbrical muscles, the superficial palmar arterial arch (**6.8.7**), and the digital vessels and nerves. A thin **intermediate palmar septum**, which usually forms an effective barrier to pus when the palm is infected, passes between the long flexor tendons of the index and middle fingers to attach to the shaft of the third metacarpal alongside adductor pollicis. This creates a **thenar space** containing the long flexor tendons of the thumb and index finger, and a **middle palmar space** containing the tendons of the middle, ring, and little fingers. Distally, these spaces extend into the sides of the fingers along the lumbricals.

Qu. 7G *By what routes might infected fluids in the thenar space and mid-palmar space be drained surgically?*

Dorsum of the hand

Finally, turn to the dorsal aspect of the hand. Review the muscles of the posterior compartment of the forearm, their position at the wrist, and the extensor retinaculum. As on the flexor aspect of the wrist, the tendons of the extensors of the fingers lie within a common synovial sheath beneath the retinaculum, but all the other tendons have separate synovial sheaths and none extends into the fingers.

Examine again the aponeurotic **extensor expansions** (**6.7.7**) which wrap around the sides of the proximal phalanges like a hood. Identify the insertions of the lumbrical and interosseous muscles to the apex of the expansions, noting that their attachments lie to the palmar side of the flexion–extension axis of the metacarpo-phalangeal joints, on either side of the deep transverse metacarpal ligament.

Requirements:

Articulated skeleton
Hammer, briefcase, pen, and key
Prosections of the hand showing palmar aponeurosis, muscles of the thenar and hypothenar eminences, long flexor tendons, adductor pollicis, interossei; flexor tendon sheaths, dorsal digital expansions
Radiographs of hand.

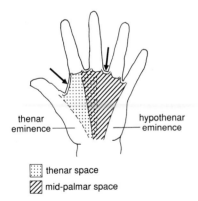

thenar eminence — hypothenar eminence

thenar space
mid-palmar space

6.7.16
Palmar fascial spaces.

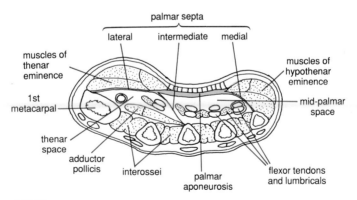

palmar septa
lateral   intermediate   medial

muscles of thenar eminence

1st metacarpal

thenar space

adductor pollicis

interossei

palmar aponeurosis

muscles of hypothenar eminence

mid-palmar space

flexor tendons and lumbricals

6.7.17
Cross-section of palm showing palmar fascial spaces

Seminar 8

Blood supply and lymphatics of upper limb

Aims To study the arterial supply and venous drainage of the upper limb; to consider the vascular anastomoses which are present around its joints; to consider the lymphatic vessels along which extracellular fluid returns to the venous circulation, and the lymph nodes along the course of these vessels.

Before you proceed, review the sections on arteries, veins, and lymphatics in Ch.2 (p. 10).

A. Living anatomy

The pulsations of the **arteries** to the upper limb can be felt at certain points along their course where they can be compressed against bone (**6.8.1**).

Stand behind your partner and feel for the pulsation of the **subclavian artery** as it crosses the 1st rib by pressing your index finger downward between the clavicle and the upper border of the scapula in the front of the neck. It is possible to compress the artery against the 1st rib if there is severe haemorrhage from the arm but this is painful because nerve trunks lie close to the artery as it enters the axilla.

Place the flat of your hand against the upper and medial aspect of the arm with your fingers as high in the apex of the axilla as possible. Feel for the pulsation of the **axillary artery** and compress it against the medial aspect of the upper end of the humerus.

Feel for the pulsation of the **brachial artery** against the mid-shaft of the humerus, and also where it lies on the medial side of the biceps tendon as it crosses the front of the elbow. It may (rarely) be necessary to compress the brachial artery in the upper arm if severe haemorrhage is occurring from the vessel distally and direct pressure has failed to stop the bleeding. This is best done by squeezing the artery against the mid-shaft of the humerus (usually by applying a tourniquet), anterior to coracobrachialis.

The brachial artery, at the front of the elbow, is a convenient site at which to measure the blood pressure. The artery is constricted in the upper arm by a pneumatic tourniquet cuff connected to a manometer; the cuff is inflated until the vessel is occluded and the radial pulse has disappeared. A stethoscope is then applied over the brachial artery just medial to the tendon of biceps and the compression of the tourniquet is slowly released until the sound of pulses of blood flowing through the artery is first heard; this indicates the systolic

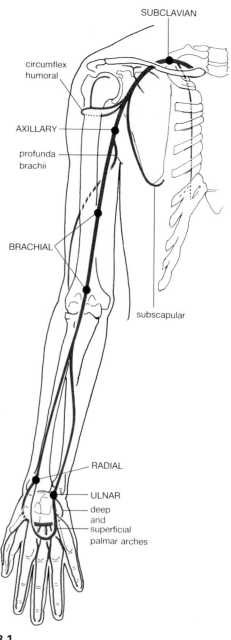

6.8.1
Major arteries of upper limb; ● pressure points for arrest of haemorrhage.

blood pressure. As the tourniquet compression is released further, the sounds at first increase and then quite suddenly disappear, when blood flow becomes continuous; this indicates the diastolic blood pressure.

Feel the pulsation of the **radial artery** at the wrist and note the consistency of its wall by compressing the artery against the flattened distal end of the radius. Count the pulsations of the artery for one minute to determine the pulse rate, and note whether the pulse is regular. Next, feel the pulsations of the radial artery as it crosses the lateral aspect of the wrist in the floor of the 'anatomical snuffbox' (p. 60). Feel also for the pulsation of the **ulnar artery** as it crosses the flexor retinaculum.

Qu. 8A *Of what do the walls of the radial artery consist?*

Examine the pattern of **superficial veins** on the back of the hand and wrist, and trace the arch on either side to the **cephalic** and **basilic veins** (**6.8.2**). With the upper arm gripped firmly to prevent venous return, palpate the cubital fossa at the front of the elbow and locate a vein into which a hypodermic needle could be inserted. This is the **median cubital vein**, which links the basilic and cephalic veins and which also receives blood from deep tissues of the forearm. The ulnar artery may be anomalously superficial in this position.

Qu. 8B *How would you decide whether what you feel is artery or vein?*

Place your right index finger on the distal end of one superficial vein on the forearm and occlude it. With the index finger of the other hand, express the blood from the proximal end of the vein. The vein is likely to remain empty. Harvey (1628) performed

this simple experiment to demonstrate that venous blood was returned to the heart through veins which contained valves.

Qu. 8C *What would be the explanation if the distal end of the vein were to fill with blood when you removed the pressure of your proximally placed left index finger?*

Using a skin-marking pencil draw on your partner the course of the major vessels of the upper limb.

Examination of the living **lymphatic system** is dealt with at the end of the seminar (p. 71).

B. Prosections

Arterial supply to the upper limb (6.8.1)

Examine the prosected part and identify the **subclavian artery** as it arises from the arch of the aorta and crosses the first rib. As the artery enters the axilla between the clavicle and outer margin of the 1st rib it is renamed the **axillary artery**. In the axilla the artery is accompanied by the brachial plexus of nerves arising in the neck and thorax (see **6.9.3**). In the natural state this neurovascular bundle is packed around with fascia and protected by fat; it also gains added protection from the muscles of the anterior and posterior walls of the axilla.

The axillary artery gives a branch (**acromio-thoracic artery**) which supplies the superficial tissues of the shoulder region anteriorly, and branches to the chest wall (**superior** and **lateral thoracic arteries**) which also supply the breast and become enlarged in lactation. It also gives

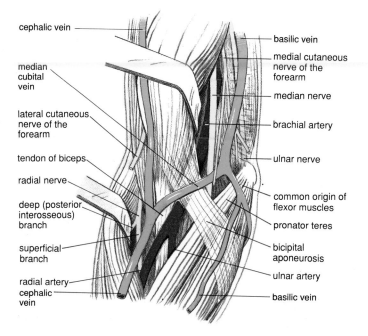

cephalic vein

median cubital vein

lateral cutaneous nerve of the forearm

tendon of biceps

radial nerve

deep (posterior interosseous) branch

superficial branch

radial artery

cephalic vein

basilic vein

medial cutaneous nerve of the forearm

median nerve

brachial artery

ulnar nerve

common origin of flexor muscles

pronator teres

bicipital aponeurosis

ulnar artery

basilic vein

6.8.2
Anterior aspect of elbow; 'cubital fossa'.

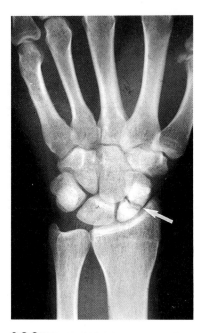

6.8.3
Fractured scaphoid.

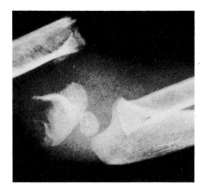

6.8.4
Supracondylar fracture of humerus
in child aged 5 (note oval
epiphysis of capitulum).

anterior and posterior **circumflex humeral arteries** which encircle the upper end of the shaft of the humerus and a **subscapular artery** which follows the lateral border of the scapula.

As it leaves the axilla at the lower border of teres major the axillary artery is renamed the **brachial artery**. Follow it down the medial aspect of the arm to the front of the elbow. In this position it is quite superficial and lies medial to the prominent tendon of the biceps, where its pulsations can be felt. During its course, the brachial artery gives rise to small branches which supply the humerus (nutrient artery) and muscles of the anterior compartment. In the axilla it gives off the **profunda brachii artery** which accompanies the radial nerve in its spiral course around the humerus to supply the muscles of the posterior compartment (triceps), and the elbow joint.

As the brachial artery passes into the forearm it divides into **radial** and **ulnar arteries** (**6.8.2**) which continue towards the wrist lying between the superficial and deep flexor muscles of the forearm. Soon after its formation the ulnar artery gives off a branch which passes deeply and divides to give two **interosseous arteries** which run distally on either side of the interosseous membrane supplying the deep tissues of the anterior and posterior compartments of the forearm. At the wrist the ulnar artery lies relatively superficially and passes into the palm on the radial side of the pisiform bone. At this point it divides into superficial and deep branches; the **superficial branch of the ulnar artery** turns laterally, crosses the long flexor tendons superficially, and anastomoses with a smaller superficial branch of the radial artery to form the **superficial palmar arch** (see **6.7.10, 6.7.11**). This arch lies at the level of the distal border of the outstretched thumb. The **deep branch of the ulnar artery** passes through the muscles of the hypothenar eminence, then deeply beneath the long tendons to anastomose with a deep branch of the radial artery to form the **deep palmar arch** which lies at the level of the proximal border of the outstretched thumb (**6.8.1** and see **6.7.11**).

As the radial artery reaches the wrist it gives off a superficial branch which crosses the muscles of the thenar eminence to anastomose with the superficial branch of the ulnar artery. The continuation of the radial artery crosses the lateral aspect of the wrist joint under cover of long tendons passing to the thumb (i.e in the anatomical snuff box). On the dorsum of the hand it passes between the adjacent heads of the 1st dorsal interosseous muscle to reach the palm deep to the long flexor tendons to the fingers; here, it anastomoses with the deep branch of the ulnar artery to form the deep palmar arch.

The fingers are supplied by **digital arteries** which arise from the superficial palmar arch. They divide opposite the metacarpal heads, and pass along the sides of the fingers to their tips, lying just dorsal to the digital nerves. Other fine arteries reach the dorsum of the digits from vessels on the dorsum of the carpus. The deep structures of the palm are supplied by branches of the deep arterial arch.

The blood supply of the **scaphoid** should be noted. As the radial artery crosses the 'snuffbox' it lies on the scaphoid to which it gives small branches which enter that bone at about its 'waist'. Some branches pass to the proximal part of the scaphoid, others to its distal pole. Falls on the outstretched hand not infrequently fracture the scaphoid across its 'waist' (**6.8.3**; arrow). Such fractures can therefore interrupt the blood supply to the proximal pole which then undergoes avascular necrosis. The proximal pole of the scaphoid articulates in the wrist joint; if it dies, movement at the wrist becomes painful and limited. It is therefore very important to check for scaphoid fractures after such falls and to immobilize the wrist to give a good chance of healing should such a fracture be suspected.

Arterial anastomoses

Muscles and joints of the upper limb are, in general, supplied by muscular branches which arise from adjacent arteries. Wherever a considerable amount of movement takes place (i.e. at joints or between parts of the body such as the scapula and the chest wall) the blood supply to the area will be plentiful and **anastomoses** between branches of nearby arteries will almost certainly be found.

On arteriograms and on prosections look for the anastomoses of arteries around the scapula (see **6.8.8**), around the elbow joint (see **6.8.6**), around the carpus and in the hand (see **6.8.7**).

Qu. 8D   *If someone had sustained a deep cut in the palm of the hand how would you stop the resulting haemorrhage?*

Qu. 8E   *Why are the walls of arteries thicker and more elastic than those of veins?*

Ischaemic contracture

Supracondylar fractures of a child's humerus (**6.8.4**) which occur as a result of a heavy fall on the outstretched arm are relatively common. One complication of such a fracture is that the flow of blood along the brachial artery may become obstructed, usually because of reflex constriction of the vessel. Although gangrene of the hand or forearm does not usually occur, muscle tissue deep in the forearm may die, depending on the efficiency of the collateral circulation at the time. The dead muscle tissue contracts by fibrosis, and this leads to a fixed flexion contracture of the wrist, extension contracture of the metacarpo-phalangeal joints, flexion contracture of the interphalangeal joints, and abduction and extension contracture of the thumb at its carpo-metacarpal joint (Volkmann's ischaemic contracture). In severe cases the median and other nerves may also be involved, producing nerve palsies (p. 81).

Venous drainage of the upper limb (6.8.5)

The venous drainage of a limb should be considered in terms of:
a) The superficial drainage of the skin and underlying superficial fascia.

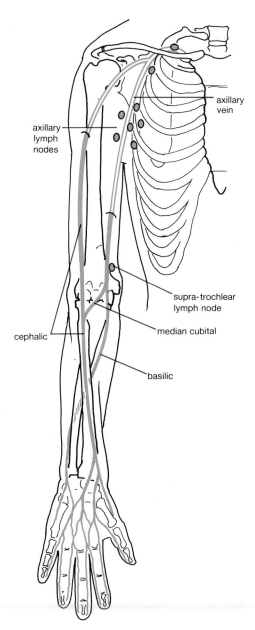

dorsal venous arch is drained by the **cephalic vein** which passes laterally up the forearm draining its superficial aspects. In the upper part of the arm it lies in the groove between pectoralis major and deltoid, a site at which venous catheters may be introduced. It then passes through the fascia which forms the anterior wall of the apex of the axilla just beneath the clavicle to drain into the axillary vein.

The pattern of minor veins is very variable but there is usually a connecting vein, the **median cubital vein**, between the cephalic and basilic veins which crosses the cubital fossa (**6.8.2**) at the front of the elbow. Because veins in the front of the elbow are firmly anchored by connective tissue to underlying structures, the cubital fossa is a convenient place to perform a venepuncture. **Deep veins** of the hand and forearm are usually found in pairs, **venae comitantes**, accompanying any small artery. These small veins eventually unite to form larger vessels which ultimately drain into the axillary vein. The axillary vein receives tributaries which correspond to the branches of the axillary artery, and becomes the **subclavian vein** at the outer border of the first rib. This joins with veins draining the head before returning blood to the heart.

During exercise or in hot weather the superficial veins of the extremities dilate and aid heat loss by radiation and convection. Under resting conditions when the ambient temperature is cool, the characteristic arrangement of the deep vessels enables heat to be distributed by counter-current mechanisms from the small arteries to their accompanying venae comitantes, thereby helping to protect the core temperature.

Lymphatic drainage of the upper limb (6.8.5, 6.8.10)

The lymphatic system is very difficult to demonstrate in the living person and in the dissecting room. **Lymph vessels** are not palpable in the living and, normally, **lymph nodes** are difficult to feel. The smaller vessels are difficult to dissect and, in the elderly, lymph nodes tend to be atrophic. The system is, however, extremely important in the spread and control of infection and malignancy.

The lymphatic drainage of the upper limb parallels that of the venous drainage in that superficial lymphatics accompany the large superficial veins, and drain the surrounding skin and superficial fascia. They receive relatively few channels from the deeper tissues. Deep lymphatics follow the deep blood vessels and drain the deeper tissues. Thus, lymphatics from the thumb and thumb web, and from the radial part of the forearm and arm, tend to drain along vessels passing with the cephalic vein; those from the ulnar side of the forearm along the basilic vein.

Both superficial and deep vessels eventually drain into the **axillary lymph nodes**. A lateral group of axillary nodes lies along the axillary artery and drains much of the lymph from the arm. Lymph passes on to central nodes lying more deeply in the axilla and thence to apical nodes in the apex of the axilla. Lymph from the anterior and

6.8.5
Superficial veins of upper limb; position of groups of lymph nodes.

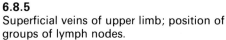

b) The deep drainage of structures which lie beneath the thick connective tissue (deep fascia) covering the muscles.

Digital veins lie on the medial and lateral aspects of the fingers and drain into a **dorsal venous arch** lying on the back of the hand. The medial (ulnar) aspect of this arch is drained by the **basilic vein** which passes proximally receiving tributaries from the medial aspect of the forearm. Above the elbow, the basilic vein pierces the deep fascia and is joined by deep veins running with the brachial artery, which have drained the deep structures of the forearm and arm. Together they form the **axillary vein** which lies medial to the axillary artery in the axilla. The lateral (radial) aspect of the

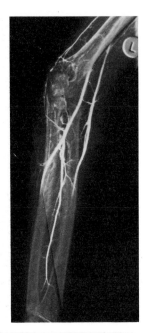

6.8.6
Arteriogram of radial and ulnar arteries.

posterior walls of the axilla drains first to local nodes and from there to more central axillary nodes. Lymph vessels running with the cephalic vein enter the apex of the axilla with the vein and end in nodes there. The apical nodes give rise to a **subclavian lymph trunk** which runs with the subclavian artery to end with other lymphatic trunks, in particular the large thoracic duct on the left side, by draining into the venous system at the confluence of the subclavian and internal jugular veins at the root of the neck. A few (supratrochlear) nodes may be situated just above the medial epicondyle.

In addition to draining the upper limb, axillary nodes also drain the skin and superficial fascia of the trunk above the umbilicus, and the breast.

Qu. 8F   *If you had a septic little finger, where would you be likely to find enlarged lymph nodes?*

With your partner's arm resting lightly at his side, try to feel pea-sized lymph nodes in the axilla. The highest (apical) nodes are quite difficult to reach with the tip of the examining finger. Normal lymph nodes often cannot be felt, but a palpable lymph node is not necessarily an indication of active pathology, because it may result from fibrosis of a previously infected node. Lymph nodes draining infected areas are often enlarged and tender, and inflamed lymphatics can be seen as reddened cords along the path of the veins. Lymph nodes are also enlarged (but less frequently tender) when involved by neoplasms which may be benign or malignant, intrinsic to the node, or secondary to spread from a primary tumour in their area of drainage.

C. Radiology

Review the section in Chapter 4 on the use of contrast media in medical imaging. **Angiograms** are made by taking radiographs after the injection of radio-opaque material into the circulation; they can be used to study arteries, veins, or lymphatics.

Arteriograms

Examine the arteriograms of the upper limb (**6.8.6–6.8.8**) and identify as many as possible of the vessels you have studied. The anastomoses around the scapula, elbow, and wrist are more easily seen on angiograms than in dissections. **6.8.8a,b** are digital subtraction angiograms showing the subclavian and axillary arteries with the arm by the side, and raised. Note how the axillary artery is bent after it has passed between the 1st rib and the clavicle; also the vessels taking part in the anastomosis around the scapula. Now turn back to **4.19** and identify the circumflex humeral vessels, the profunda brachii, radial, ulnar, and interosseous arteries; note too the anastomoses around the joints.

The following case illustrates how a knowledge of the arterial supply to the arm enabled a surgeon to trace and remove an acute blockage in a major vessel. A car driver was severely injured in a head-on collision with a lorry. Before operation it was observed that the right arm was abnormally cold. No pulse could be felt in either the right wrist or the right axilla, but one was present in the left arm. Arteriograms were taken to locate the site of blockage of the circulation in his right arm.

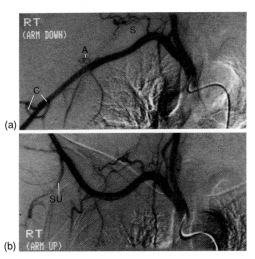

6.8.8
Digital subtraction angiograms of subclavian and axillary arteries with (a) arm hanging down, (b) arm raised vertically above the head. S—subclavian artery; A—acromio-thoracic artery; C—circumflex humeral arteries; SU—subclavian artery.

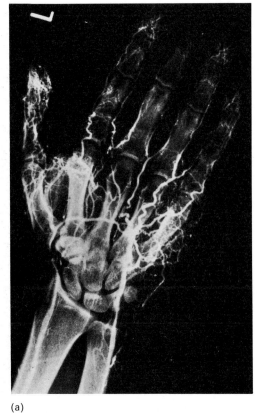

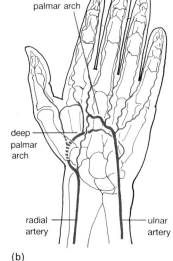

superficial palmar arch

deep palmar arch

radial artery   ulnar artery

(a)   (b)

6.8.7
(a) Arteriogram of palmar arches and digital supply. (b) Labelled diagram.

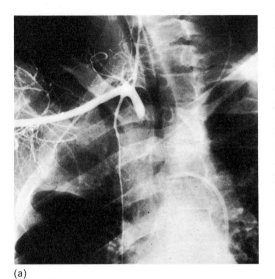

(a)

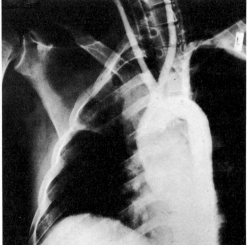

(b)

6.8.9
See Qu. 8G.

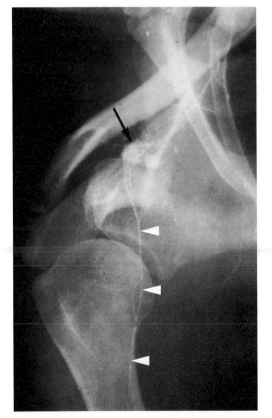

6.8.10
Lymphogram of vessels (arrowheads) and
nodes (arrow) of axilla.

Through a small incision near the elbow a plastic
catheter was passed retrogradely (i.e. against the
flow of blood) up the right brachial artery, to the
axillary artery. Another catheter was passed retrogradely up the right femoral artery to the proximal
end of the arch of the aorta. When the catheters
were in place, radio-opaque medium was injected
first into the axillary artery and a radiograph taken
(**6.8.9a**). Opaque medium was then injected into the
proximal end of the aortic arch and further
radiographs were taken (**6.8.9b**).

Qu. 8G *Examine Figs. **6.8.9a & b.** Where was the
site of arterial obstruction in this patient?*

Lymphograms (**6.8.10**) can be prepared by
taking radiographs after injection of radio-opaque
material into small lymphatic vessels which have
previously been made visible by the subcutaneous
injection of patent blue dye which is taken up by the
lymphatics. Lymphangiograms can be used to
detect larger vessels, lymph nodes, and sites of
blockage of lymphatic drainage. Lymph nodes can
also be detected by the injection of radioactive
colloids which are phagocytosed by macrophages
which sequester in the nodes.

Requirements:

Prosections of arterial supply to the upper limb
Prosections of major superficial and deep veins
and their entry into the axillary vein
Angiograms and lymphograms of the upper limb
Skin marking pencils (red & blue)
Sphygmomanometer and stethoscope.

Seminar 9

Innervation of upper limb: brachial plexus

Aims To consider the principles which underlie the innervation of the upper limb; to study the brachial plexus by which motor and sensory nerves of lower cervical and upper thoracic spinal neurons are distributed to muscle groups, joints, and skin of the upper limb; and to consider the disabilities which arise if parts of the plexus are damaged. To consider the origin and functions of the autonomic sympathetic nerves which supply blood vessels and sweat glands.

Before you proceed, review the section on innervation in Chapter 2 (p. 11).

A. Dissection and prosections

Now that you are familiar with prosections of the muscles, joints, and blood supply, it is useful to dissect the axilla, identify the components of the brachial plexus, and follow the course and distribution of its main branches. In this way you will also review the muscles and become aware of the relationships between the nerves, arteries, and muscles.

Blunt dissection, using your fingers or a pair of forceps, or by opening a pair of scissors in an anatomical space is nearly always to be preferred to the use of a scalpel since it will enable tissues to be cleaved along anatomical planes with minimal risk of destroying important arteries or nerves. A scalpel should be used only when you are certain about what you are cutting.

If you have not already done so, make the skin incisions illustrated in **6.2.2** and reflect the whole skin, i.e. the epidermis and dermis (including the mammary gland), from the underlying muscles by inserting your fingers or the handle of a pair of forceps between the dermis and the deep fascia overlying the muscles. When your fingers are cleaving the plane between dermis and muscles, you will feel the resistance of certain cutaneous nerves which pass through the subcutaneous tissue to supply the skin. Follow these nerves through the subcutaneous tissue as far as possible before cutting them.

Having reflected the skin of the chest and upper arm, define the borders of pectoralis major (p. 40), remove its covering of deep fascia, slide your fingers deep to it near its insertion, and cut through it with a scalpel. Reflect the muscle towards the sternum and then reflect pectoralis minor in the same way after cutting it 1–2 cm from its insertion into the

coracoid process. The anterior wall of the axilla will now have been reflected to expose the axillary neurovascular bundle surrounded by fat.

Free the mid part of the clavicle from attached tissue, scrape away the periosteum, and cut away the middle of the clavicle using a bone saw and bone chisel, taking care not to damage structures lying deep to the clavicle. This will expose the neurovascular bundle as it crosses the 1st rib. Now remove the fat from the axilla to reveal its contents.

Exposure of the mid-part of the 1st rib reveals the lower attachment of a neck muscle, **scalenus anterior**. This muscle, which is better seen in a prosection of the neck (Vol. 3), arises from the anterior aspect of the transverse processes of cervical vertebrae C3–6. The muscle is easily recognized because the subclavian vein lies anterior to it as it crosses the first rib, while the subclavian artery and the nerve roots lie posterior to it. Behind scalenus anterior, the subclavian artery and the nerve roots, lie **scalenus medius** and **scalenus posterior**. These muscles arise from the posterior aspect of the transverse processes of the cervical vertebrae and insert into the outer aspects of the 1st and 2nd ribs. The scalene muscles therefore lie on either side of the intervertebral foramina through which the cervical spinal nerves C5 to T1 emerge. (Remember that there are 8 cervical spinal nerves but only 7 cervical vertebrae.)

The formation and pattern of the roots, trunks, and cords of the plexus are easy to identify, but difficult to retain in the memory for any length of time. There is little to be gained from a detailed knowledge of this branching which, you will see, varies somewhat from body to body. Important aspects of the plexus, which you should remember, are:

- The relationships of the roots of the plexus to the intervertebral foramina.

- The position of the plexus on the side of the neck.

- The general motor distribution of the roots of the plexus to functional groups of muscles.

- The roots involved in the reflex 'jerks' elicited during clinical examination.

- The general sensory distribution of the roots of the plexus.

- The major nerves arising from the plexus.

The relationships of the roots of the plexus to the intervertebral foramina are dealt with in the seminar on the spine (**8.29**). Briefly, however, the **ventral** and **dorsal spinal nerve roots** emerge from the spinal cord within the spinal canal, and unite in an intervertebral foramen to form a **spinal nerve**. The **dorsal root ganglion**, containing cell bodies of the peripheral sensory fibres, is usually situated in the intervertebral foramen. Immediately distal to the foramen the spinal nerve divides into **anterior** and **posterior primary rami**. The anterior primary rami form the **roots of the brachial plexus**; the posterior primary rami supply extensor muscles of the spine and skin of the back.

A knowledge of the position of the brachial plexus in the neck is important for any anaesthetist who is required to 'block' nerves or inject radio-opaque material into arteries in the vicinity.

Extend your neck and bend it laterally to one side. If you now depress your shoulder on the opposite side, you will be able to feel the trunks of the brachial plexus as taut bands immediately above the clavicle.

The **roots of the brachial plexus (6.9.1, 6.9.2)** are formed from the anterior primary rami of C5–T1 spinal nerves. Locate them between scalenus anterior and scalenus medius and clean them of connective tissue. Note that T1 root has to emerge from the chest cavity.

From the roots, branches pass to supply the scalene muscles (segmentally); subclavius (C5 & 6); the rhomboid muscles via a nerve (C5) that passes dorsally through levator scapulae to reach the back; and serratus anterior via the nerve to serratus

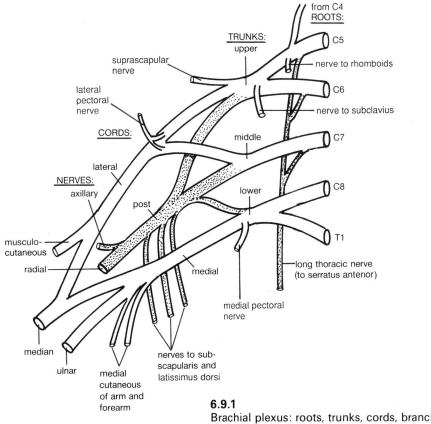

6.9.1
Brachial plexus: roots, trunks, cords, branches.

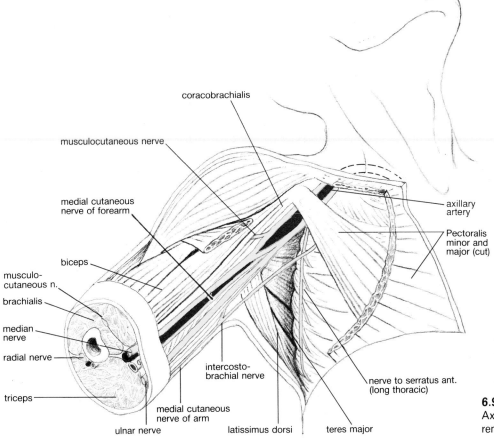

6.9.2
Axilla; most of pectoralis major has been removed to expose the neurovascular bundle.

Table 1. Segmental nerve supply to movements of the upper limb

Joint	Muscle action	Root supply	Muscle action	Root supply
Shoulder	Abduction Lateral rotation	C5	Adduction Medial rotation	C6, 7, 8
Elbow	Flexion	C5, 6	Extension	C 7, 8
Radio–ulnar	Supination	C 6	Pronation	C 7, 8
Wrist	Flexion and Extension	C6, 7		
Fingers	Long Flexors and Extensors	C7, 8		
Hand	Small muscles	T1		

anterior or **long thoracic nerve** (C5, 6, 7) which lies on the side of the chest wall and is vulnerable in operations for removal of the breast. Locate this nerve as it crosses the first rib deep to the axillary artery and follow it as it descends and supplies serratus anterior on the side of the chest.

Qu. 9A *What action would be lost if the long thoracic nerve were damaged?*

The **trunks of the brachial plexus** are formed from the union of its roots. C5 and C6 roots join to form the **upper trunk**; C7 root continues alone to form the **middle trunk**; and roots C8 and T1 join to form the **lower trunk**. The trunks lie relatively exposed on the side of the neck.

Trace the **suprascapular nerve** (C5, 6) which arises from the upper trunk and passes backward across the lower part of the posterior triangle to reach the suprascapular notch through which it passes to supply supraspinatus and infraspinatus.

Qu. 9B *What actions would be lost if the suprascapular nerve were damaged?*

Nerve bundles of the upper, middle, and lower trunks of the plexus divide and are redistributed beneath the clavicle into **anterior divisions** and **posterior divisions** (which supply, respectively, flexor and extensor musculature and skin; p. 28) which unite to form the **cords of the brachial plexus**.

There are three **cords** which are named according to their position relative to the mid-part of the axillary artery. Identify the **lateral cord** and its branches: a branch to pectoralis major (lateral pectoral nerve); the musculocutaneous nerve which passes into coracobrachialis; and a contribution to the median nerve.

Identify the **medial cord** and its branches which pass between the axillary artery and vein: a branch which supplies pectoralis major and minor (medial pectoral nerve); the small medial cutaneous nerve of the arm; the larger medial cutaneous nerve of the forearm; a contribution to the median nerve; and the ulnar nerve.

Tie off and cut the axillary vein as it crosses the 1st rib and remove it together with its tributaries in the axilla. Pull the axillary artery laterally to reveal the **posterior cord** of the brachial plexus and its branches: branches which supply the muscles of the

posterior wall of the axilla (subscapularis, teres major, and latissimus dorsi); the axillary or circumflex nerve, which passes backward to leave the axilla immediately inferior to the shoulder joint between subscapularis and teres major; and the radial nerve which runs downward through the axilla to pass between the long and medial heads of triceps into the posterior compartment of the arm.

Qu. 9C *What is the function of a nerve plexus?*

The motor distribution of the roots

Each upper limb muscle has a nerve supply derived from one or more segments of the spinal cord. Memorizing root values of the nerve supply to individual muscles is unnecessary; they can be looked up in a reference textbook. Moreover, they can be determined by remembering certain broad principles, because muscles are innervated according to the movement they produce. Also, the movement of a joint is more easily tested than the action of particular muscles so that, by analysing which movements are impaired by paralysis after an injury, the level of neurological involvement of the plexus or spinal cord can be determined.

Each spinal nerve supplies certain functional groups of muscles. Table 1 indicates the nerve roots which are responsible for particular movements of the upper limb. There is a pattern of arrangement to the supply, but it is quite different from that of the supply to the skin. In general, more distal muscle groups are supplied by more caudal spinal nerves; opposite movements of a joint are supplied by adjacent spinal segments.

The 'reflex jerks' of the arm are stretch reflexes produced by tapping muscle tendons. A reflex jerk depends upon the integrity of (a) afferent nerves from tendon and muscle receptors to the spinal cord, (b) efferent motor pathways from the cord to the muscles concerned, and (c) the degree of excitation of motor neurons in the segment of the spinal cord at which the reflex connections occur. Thus reflexes should always be compared on the two sides of the body. The biceps jerk and the triceps jerk are the two most commonly examined; their root value can be found in the Table above. Analysis of reflexes is an important part of clinical assessment of the nervous system.

The sensory distribution of the roots

The sensory distribution of the roots provides another method of determining the level of a neurological lesion. The pattern of loss of sensation in the arm may correspond to that of a major nerve, or of a larger part of the plexus. The area of skin supplied by one dorsal root of the spinal cord is known as a sensory **dermatome**. Maps of the skin of the arm, as for the rest of the body, have been prepared on the bases of clinical cases of nerve root damage. **6.9.3** is a diagram of such a map.

Again, general principles are important:

- progressively more caudal spinal segments supply the preaxial border of the limb, the digits, and the postaxial border (**6.9.3**);

- the middle nerve of the plexus (C7) supplies the middle digit;

- overlap of supply is least between non-adjacent (C5 and T1, C6 and C8) segments, but is considerable between territories supplied by adjacent spinal nerves.

Furthermore, after injury, the supply from intact dermatomes may invade the periphery of anaesthetic areas. Thus, when one spinal nerve is lesioned, a neatly defined area of anaesthesia suggested by **6.9.3** is *not* produced.

The lines between non-adjacent dermatomes are referred to as axial lines and are more pronounced on the flexor than the extensor surfaces of the limbs. This is made use of in clinical assessment of patients for a 'sensory level' of a spinal nerve lesion by testing changing sensory *ability* across an axial line.

The **autonomic nerve** supply to vascular smooth muscle, sweat glands and arrector pili muscles (attached to skin hairs) of the upper limbs consists of postganglionic **sympathetic** fibres only. The preganglionic cell bodies serving the arm lie in the upper thoracic segments of the spinal cord in a lateral horn of grey matter; their myelinated fibres (white rami communicantes) pass into the ventral spinal nerve roots but leave them just outside the vertebral column to enter the sympathetic chain—a system of interconnected ganglia and nerve fibres lying vertically on either side of the vertebral column. The terminal arborizations of the white rami synapse on ganglion cells in the inferior cervical ganglion or first thoracic ganglion (which may be fused to form a 'stellate' ganglion) or in the middle cervical ganglion (Vol. 3, p. 130). Unmyelinated postganglionic fibres (grey rami communicantes) mostly rejoin the ventral roots and are distributed with the branches of the brachial plexus, though some pass directly on to the subclavian artery and are distributed via the arterial tree.

Damage to the brachial plexus

By combining anatomical knowledge with the findings on examination of sensory loss, muscle paralysis, and reflex loss, a clinician can usually determine the level of a neurological lesion.

Most injuries to the brachial plexus occur as a result of severe traction. A fall from a motorcycle at speed is a common source of injury. The rider hurtles through the air and lands on the side of his head and shoulder, thereby stretching the upper roots of the plexus. Or, if his shoulder strikes some obstacle such as a car, the nerves roots may tear in sequence from above downward. The injury may also occur during childbirth if too much traction is applied to the head while the shoulder is retained in the birth canal. The lower part of the brachial plexus can be injured in the opposite way, by sudden powerful traction of the arm above the head, as for instance when someone breaks a fall from the platform of a bus by holding on to the bar, or in falling through a trap-door.

Suppose only the **upper trunk** of the brachial plexus is involved? Muscles innervated from C5 and C6 will be paralysed and eventually waste. The most important are the clavicular head of pectoralis major, deltoid, supra- and infraspinatus, coracobrachialis, brachialis, biceps, brachioradialis, supinator and extensor carpi radialis longus. The arm hangs by the side, the elbow cannot be flexed, the arm is medially rotated, and the forearm is pronated. This appearance has been called the 'waiter's tip' or 'crafty smoke' position (Erb's palsy). There will also be loss of sensation over the deltoid (C5) and lateral aspect of the forearm (C6).

Suppose the **middle trunk** (C7) of the brachial plexus is involved? Usually it is damaged in addition to the upper roots and the following muscles are also paralysed: serratus anterior, latissimus dorsi, teres major, triceps, the middle fibres of pectoralis major, pronator teres, flexor carpi radialis, flexor digitorum superficialis, extensor carpi radialis longus and brevis, extensor digitorum, and extensor digiti minimi. There is 'winging' of the scapula (**6.9.4**), loss of extension of the elbow, and radial deviation of the wrist with some weakness of power grip due to loss of extension of the wrist and fingers.

Qu. 9D   *What other disabilities would you expect to be able to elicit?*

Suppose only the **lower trunk** of the brachial plexus is injured? Muscles innervated from C8 and T1 will be paralysed. These are: all the muscles of

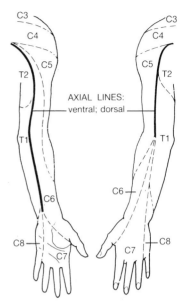

6.9.3
Dermatomes of upper limb; note the axial lines.

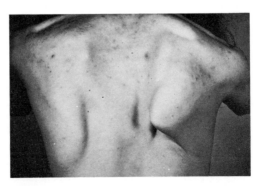

6.9.4
'Winged scapula' demonstrated by pushing forward against resistance.

the flexor compartment of the forearm (except pronator teres and flexor carpi radialis), and the intrinsic muscles of the hand. The paralysis is much the same as would be produced if the median and ulnar nerves were divided; the grip is lost. In the longer term, the muscles will waste and a claw-like deformity will result (p. 83). The skin of the little finger and the ulnar side of the palm and forearm is insensitive. If T1 and T2 nerve roots are injured, this will cut off the sympathetic supply not only to the upper limb, but also to the head and neck on the side of the lesion, resulting in a 'Horner's syndrome' (Vol. 3, p. 131).

If the **whole brachial plexus** is injured the arm hangs uselessly from the shoulder, and is entirely insensitive except for the skin over the upper part of the shoulder (supplied by fibres from supraclavicular branches of the cervical plexus; C4) and the medial aspect of the upper arm (supplied by the intercosto-brachial nerve; T2). Such an injury defies treatment. Nowadays, however, most paralysed arms are supported by splints which can be manoeuvred by the patient.

Qu. 9E *What other disabilities would you expect to be able to elicit if the entire plexus were injured?*

Do remember that it is not necessary for these lists of facts to be committed to memory. Table 1 is provided so that you can work out which spinal nerve root supplies which muscle.

Occasionally the brachial plexus may be **pre-** or **post-fixed** so that the entire plexus arises from a higher or lower set of segments. If the latter should occur, or if a **cervical rib** or fibrous band develops in the presence of a normally-derived brachial plexus, then pressure may be exerted on the lower trunk of the brachial plexus. Since the lower trunk contains motor fibres which supply the small muscles of the hand (through the median or ulnar nerves), damage to the root will cause wasting and weakness of the small muscles of the hand. Sensory changes will also be found on the medial aspect of the forearm and hand. In addition, the autonomic fibres to the blood vessels of the upper limb may be irritated or damaged, causing vascular changes which will be noticed predominantly in the hand.

Requirements:

Articulated skeleton
Prosections of the neck and axilla showing the components of the brachial plexus: roots, trunks, divisions, and branches
Bone-cutting tools.

Seminar 10

Innervation of upper limb: lateral cord distribution

Aims To study the course and distribution of the motor and sensory branches of the lateral cord of the brachial plexus; and to consider the disabilities which arise if the lateral cord or any of its branches is damaged.

A. Dissection and prosections

Locate the lateral cord on the lateral aspect of the axillary artery. Define its origin from anterior divisions of the upper and middle trunks, and trace it distally to discover its branches: the **lateral pectoral nerve** to pectoralis major; the musculo-cutaneous nerve; and a contribution to the median nerve.

The **musculocutaneous nerve (6.10.1, 6.10.2)** supplies the muscles of the flexor (anterior) compartment of the arm and the skin over the lateral aspect of the forearm.

Trace the nerve as it leaves the axilla and supplies and then pierces coracobrachialis. Find its

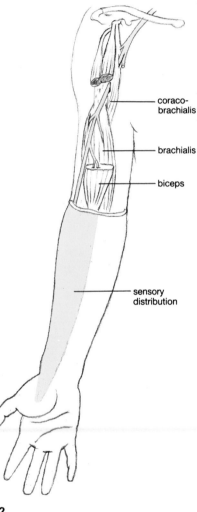

6.10.2
Course of musculocutaneous nerve; supply to skin.

branches to biceps and brachialis and then follow the nerve deep to biceps until it emerges on its lateral aspect just above the elbow. This terminal branch is the **lateral cutaneous nerve of the forearm** which supplies the skin down to the thenar eminence. You should have located and followed it, at least in part, when you reflected the skin.

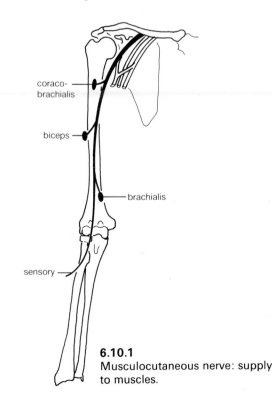

6.10.1
Musculocutaneous nerve: supply to muscles.

Qu. 10A *What disabilities would result from damage to the musculocutaneous nerve as it leaves the lateral cord?*

The **median nerve (6.10.3, 6.10.4)** is formed by contributions from both the lateral and medial cords. It supplies most of the muscles of the flexor (anterior) compartment of the forearm (but not flexor carpi ulnaris or the ulnar half of flexor digitorum profundus); the lateral two lumbricals; and the muscles of the thenar eminence. It also supplies skin over the palmar surface and the nail beds of the lateral (radial) $3\frac{1}{2}$ (or sometimes $2\frac{1}{2}$) digits.

Trace the nerve as it crosses in front of (or occasionally behind) the axillary artery to pass distally under cover of biceps through the forearm. At the elbow it lies medial to the brachial artery and therefore medial to the tendon of biceps. After crossing the anterior aspect of the elbow joint it passes into the forearm between the two heads of origin of pronator teres. As it does so it gives off a **deep branch** (anterior interosseous nerve) which passes with the interosseous branch of the ulnar artery to the deep muscles of the forearm. The median nerve is well protected in this part of the upper limb, lying between the superficial and deep

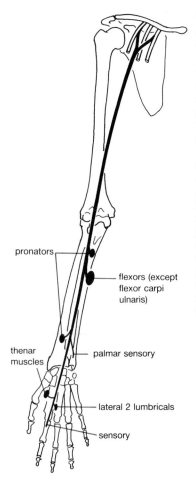

6.10.3
Median nerve: supply to muscles.

flexor muscles of the forearm; it continues distally, bound to the deep aspect of flexor digitorum superficialis by fascia, until it reaches the wrist. Before it reaches the wrist, it gives off a slender **palmar cutaneous branch** which passes over the flexor retinaculum to supply the palmar skin.

As the median nerve crosses the wrist joint it lies between flexor carpi radialis and the lateral side of the long tendons of the superficial flexor muscles. The nerve is quite superficial at this point, but immediately enters the carpal tunnel.

Trace the median nerve through the carpal tunnel and note the short, thick **recurrent branch** which leaves the median nerve as it emerges from the tunnel to double back and supply the muscles of the thenar eminence. Lastly, identify the **digital branches** to the lateral (radial) $3\frac{1}{2}$ digits and to the lateral two lumbricals; trace these branches into one finger and note that they accompany the digital arteries which pass along the sides of the fingers. The digital nerves supply both palmar and dorsal aspects of skin over the terminal phalanx, and thus the nail bed.

Qu. 10B   *If a finger needs to be anaesthetized where would you inject the local anaesthetic? Is there a possible danger in this procedure, and if so what is it?*

Review the **flexor retinaculum** and the **carpal tunnel** (p. 61). There is no room for expansion within the carpal tunnel so that, if the capsule of the carpal joints or the synovial sheaths of the flexor tendons swell, the median nerve is likely to become compressed (carpal tunnel syndrome). Such changes can occur in pregnancy when water retention occurs, in rheumatoid arthritis, in repetitive strain injury, and often for no obvious reason! If symptoms persist the retinaculum can be surgically divided.

If the median nerve is divided at the wrist, or if its compression in the carpal tunnel is sufficiently severe, there will follow:

● On **visual examination**—if the nerve damage has been present for some time, the superficial thenar muscles will have wasted, and the thenar eminence will appear 'flat'. Because of the paralysis, the thumb will be lying in the same plane as the palm and the other digits. This appearance is often described as simian (monkey-like) (**6.10.5**).

● **Motor loss**—loss of power in the thenar eminence muscles resulting in weak opposition and therefore a weak pinch grip (p. 63). If median nerve dysfunction is permanent, opposition can be restored by surgical re-routing of a tendon of flexor digitorum superficialis to the lateral border of the thumb metacarpal. The lumbricals of the middle and index fingers will also be paralysed. (The effects may vary from person to person because some muscles can be supplied from the ulnar nerve.)

● **Sensory loss**—loss of sensation over the thenar eminence and flexor aspects of the radial $3\frac{1}{2}$ digits, extending over the tips of the digits to the nail beds (**6.10.4**). This, combined with the motor loss, will prevent the ability to discriminate between small objects by touch. Early stages of

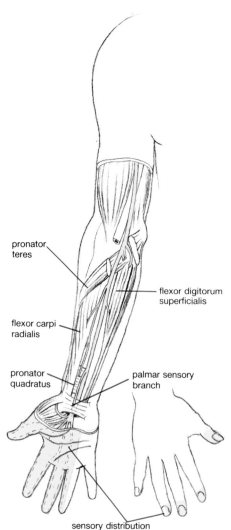

6.10.4
Course of median nerve; supply to skin.

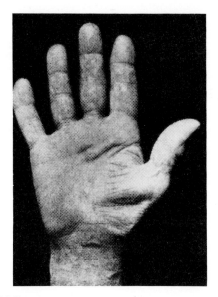

6.10.5
Right median nerve paralysis: note wasting of thenar eminence and position of thumb.

median nerve compression in the carpal tunnel are characterized by tingling sensations in the thumb and index and middle fingers which may extend to the forearm and which are especially troublesome in the early hours of the morning.

Qu. 10C *If the median nerve is damaged at the elbow or above what will be the additional effects to those mentioned above?*

Requirements:

Articulated skeleton
Prosections of the axilla, arm, forearm, and hand showing the lateral cord of the brachial plexus and the distribution of the musculocutaneous and median nerves.

Seminar 11

Innervation of upper limb: medial cord distribution

Aims To study the course and distribution of the motor and sensory branches of the medial cord of the brachial plexus; and to consider the disabilities which arise if the medial cord or any of its branches is damaged.

A. Dissection and prosections

Locate the medial cord lying between the axillary artery and axillary vein. Define its origin (largely) from the anterior division of the lower trunk, and trace it distally to discover its branches.

Locate the small **medial pectoral nerve** passing into pectoralis major to supply the pectoral muscles. Identify also the small **medial cutaneous nerve of the arm** which supplies the skin on the medial aspect of the arm; and the larger **medial cutaneous nerve of the forearm** which passes distally to supply skin on the medial aspect of the upper limb from above the elbow to the wrist. Note again the contribution of the medial cord to the **median nerve** (p. 80).

The main continuation of the medial cord is the **ulnar nerve** (**6.11.1**, **6.11.2**, see **6.8.2**). This nerve supplies flexor carpi ulnaris; the ulnar half of flexor digitorum profundus; the small muscles of the hand with the exception of those supplied by the median nerve, and the skin of both the palmar and dorsal aspects of the ulnar (medial) $1\frac{1}{2}$ (or $2\frac{1}{2}$) fingers.

Trace the ulnar nerve, which runs distally medial to the axillary artery and its continuation the brachial artery, and then on to reach the posterior aspect of the medial epicondyle. Note that the nerve lies first in the anterior compartment of the arm and then pierces the medial intermuscular fibrous septum to enter the posterior compartment where it lies anterior to triceps. It then passes behind the medial epicondyle where it can be palpated in the living body and can easily be injured. Hitting or bruising the nerve here can cause a burning or tingling sensation which runs down the forearm to the little and ring fingers—hence the layman's term of 'funny bone' for the bone on the medial side of the elbow! Follow the nerve distally from the medial epicondyle, noting that it supplies and then pierces flexor carpi ulnaris to reach the anterior compartment of the forearm. In the forearm, the ulnar nerve continues distally under flexor carpi ulnaris, lying on flexor digitorum profundus to which it gives a branch. As it nears the wrist a **dorsal cutaneous branch** is given off which passes on to the

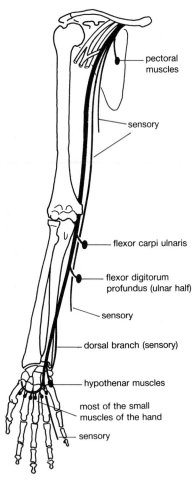

6.11.1
Ulnar nerve: supply to muscles.

Labels on figure 6.11.1:
- pectoral muscles
- sensory
- flexor carpi ulnaris
- flexor digitorum profundus (ulnar half)
- sensory
- dorsal branch (sensory)
- hypothenar muscles
- most of the small muscles of the hand
- sensory

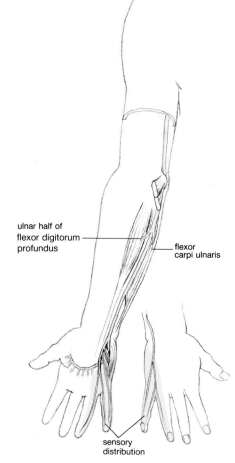

6.11.2
Course of ulnar nerve; supply to skin.

Labels on figure 6.11.2:
- ulnar half of flexor digitorum profundus
- flexor carpi ulnaris
- sensory distribution

posterior aspect of the wrist to supply the dorsal aspect of the medial (ulnar) $1\frac{1}{2}$ (or $2\frac{1}{2}$) fingers. As the ulnar nerve crosses the wrist, on the radial side of the tendon of flexor carpi ulnaris, it lies in a small fibrous tunnel superficial to the main part of the flexor retinaculum. When it reaches the muscles of the hypothenar eminence it divides into superficial and deep branches. The **superficial branches** can be felt in the living body by palpating the hook of the hamate on the hypothenar aspect of the palm. Roll the pad of your thumb firmly over the prominent hook and you should be able to feel the cord-like superficial branch of the ulnar nerve as it passes across the bone. The superficial branches form digital nerves which pass on either side of the little finger and on the ulnar aspect of the ring finger to supply the skin. The **deep branch** of the ulnar nerve passes through the hypothenar eminence muscles and supplies them. Follow it as it turns laterally beneath the long flexor tendons to the fingers to lie with the deep palmar arterial arch; it supplies most of the small muscles of the hand: i.e. the dorsal and palmar interossei, the medial two lumbricals, and adductor pollicis.

To trace the deep branch of the ulnar nerve satisfactorily, you will need to cut through the flexor retinaculum and either displace or cut through the long tendons to the fingers.

Though it cannot readily be demonstrated by dissection, it is important to note that digital branches of the median and ulnar nerves to the fingers supply not only the palmar aspect of the fingers, but also the nail bed via branches that pass dorsally around the distal phalanx.

The ulnar nerve can suffer gradual traction and compression as a result of an increase in the carrying angle at the elbow due to unequal growth of the epiphyseal plate during growth. If the angle is increased (**6.11.4**) the position of the elbow which results is known as a 'valgus' (bent-outward) deformity (the opposite is 'varus'). Both valgus and varus deformities can also result from badly set supracondylar fractures.

If the ulnar nerve were damaged at the elbow there would follow:

● **Visual examination**—the hand would appear 'clawed' (**6.11.4**, **6.11.5**) (see below). If muscle wasting had occurred the hypothenar eminence would be flattened and the intermetacar-pal spaces would appear hollowed on the dorsum of the hand.

● **Motor loss**—Adduction at the wrist would be weak owing to paralysis of flexor carpi ulnaris. Flexion of the little and ring fingers would be weak and the power of abduction and adduction of the fingers would be lost. This is tested by getting the patient to try to retain a grip on a piece of paper held between extended fingers, or to spread the fingers against resistance. Because the interossei and lumbricals flex the metacarpo-phalangeal joints but extend the interphalangeal joints (p. 66) the action of the long extensor tendons is unopposed and the metacarpo-phalangeal joints become hyperextended. The interphalangeal joints would, however be flexed (the medial two more than the lateral two).

● **Sensory loss**—Loss of sensation over the palmar and dorsal aspects of the ulnar $1\frac{1}{2}$ (or $2\frac{1}{2}$) fingers.

Qu. 11A   *Why are the middle and index fingers not similarly clawed?*

Qu. 11B   *How would the results of severance of the ulnar nerve at the wrist differ from severance at the elbow?*

Requirements:

Articulated skeleton
Prosections of axilla, arm, forearm, and hand showing the medial cord of the brachial plexus and its branches: medial pectoral nerve; medial cutaneous nerves of arm and forearm; ulnar nerve; and the contribution to the median nerve.

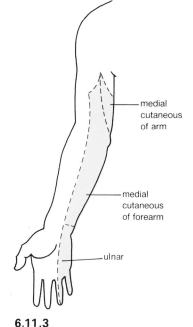

6.11.3
Distribution of medial cutaneous nerves of arm and forearm, and of ulnar nerve.

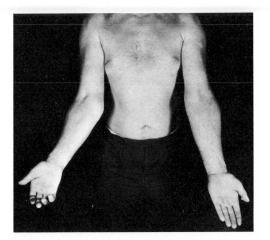

6.11.4
Increased carrying angle ('valgus deformity') of right elbow associated with ulnar nerve damage. Note 'claw' position of fingers on affected side.

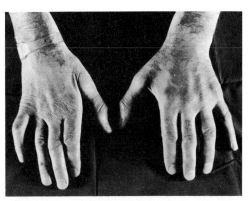

6.11.5
Right ulnar nerve palsy causing 'clawed' hand—note hyperextended metacarpo-phalangeal joints especially of ring and little fingers.

Seminar 12

Innervation of upper limb: posterior cord distribution

Aims To study the course and distribution of the motor and sensory branches of the posterior cord of the brachial plexus; and to consider the disabilities which arise if the posterior cord or any of its branches is damaged.

A. Dissection and prosections

Locate the posterior cord (**6.12.1**, see **6.9.1**) which lies immediately posterior to the axillary artery and define its origin from the posterior divisions of all three trunks. Identify its branches of supply to the muscles of the posterior wall of the axilla: the **upper subscapular nerve**, which is the most proximal of the three; the **lower subscapular nerve**, which supplies both subscapularis and teres major; and the **nerve to latissimus dorsi**, which lies between them.

Qu. 12A *Which nerves supply the muscles of the anterior and medial walls of the axilla?*

Now locate the two terminal branches of the posterior cord, the **axillary nerve** and the **radial nerve**.

Trace the **axillary nerve** (**6.12.2**) as it passes dorsally out of the axilla between the lower border of subscapularis and the upper border of teres major. As it does so, it supplies a branch to teres minor and then divides into two terminal branches. The deeper branch passes laterally around the 'surgical neck' of the humerus and supplies deltoid from its deep aspect. Cut through deltoid to expose the nerve. The more superficial terminal branch also supplies deltoid, and a small area of skin over the insertion of deltoid (**6.12.3**) which it reaches by passing around the posterior border of the muscle.

When tracing the axillary nerve through the posterior wall of the axilla you should have noticed its close proximity to the weakly supported inferior aspect of the shoulder joint. Pass your index finger

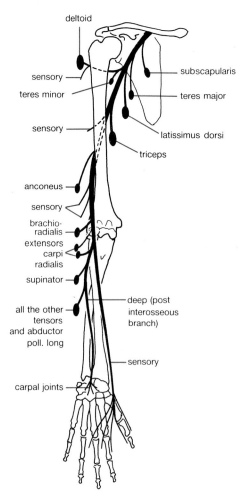

6.12.1
Axillary and radial nerves: supply to muscles.

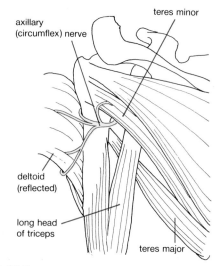

6.12.2
Course of axillary nerve.

6.12.3
Axillary nerve: supply to skin.

along the course of the axillary nerve as it leaves the axilla and press upward and feel the inferior aspect of the shoulder joint. Remember that the head of the humerus usually dislocates in an inferior (and anterior) direction, so that the axillary nerve is often bruised if not permanently damaged.

The **radial nerve** (6.12.4, 6.12.5) supplies virtually all the muscles and skin over the whole of the posterior aspect of the upper limb.

Trace the nerve as it leaves the axilla and passes backward between the long and medial head of triceps into the posterior compartment of the arm. Look for the small **posterior cutaneous nerve of the arm** which leaves the main nerve at this point to supply skin of the back of the arm. Now cut through the lateral head of triceps to locate the radial nerve as it spirals laterally around the shaft of the humerus in the radial groove, lying superficial to the uppermost fibres of the medial head of triceps; these muscle fibres can protect the nerve from injury if the bone is fractured at this level.

The radial nerve supplies branches to the three heads of triceps and a small muscle, anconeus. As the nerve approaches the lateral intermuscular fibrous septum—which it pierces to enter the anterior compartment—it gives off a small cutaneous branch which supplies the lower lateral aspect of the arm, the **lower lateral cutaneous nerve of the arm**, and a larger cutaneous branch, the **posterior cutaneous nerve of the forearm**, which passes distally to supply the skin of the posterior aspect of the forearm. After the radial nerve has entered the anterior compartment of the arm it supplies the muscles arising from the lateral supracondylar ridge of the humerus (i.e. brachioradialis and extensor carpi radialis longus). It lies under cover of these muscles on the lateral side of the elbow and is thus well protected.

Follow the radial nerve distally on to the anterior aspect of supinator, a muscle which wraps around the head of the radius. At this point the radial nerve gives off a large and important branch, the **deep branch of the radial**, or **posterior interosseous nerve**. This nerve supplies superficial extensors arising from the common extensor origin, pierces supinator, and winds laterally around the neck of the radius to reach the posterior compartment of the forearm. Here, it supplies all the remaining forearm extensor muscles and supinator. Trace this branch into the posterior compartment of the forearm by dividing brachioradialis and

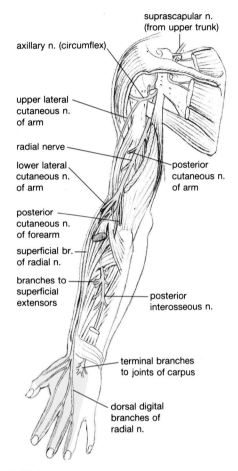

suprascapular n.
(from upper trunk)

axillary n. (circumflex)

upper lateral
cutaneous n.
of arm

radial nerve

lower lateral
cutaneous n.
of arm

posterior
cutaneous n.
of arm

posterior
cutaneous n.
of forearm

superficial br.
of radial n.

branches to
superficial
extensors

posterior
interosseous n.

terminal branches
to joints of carpus

dorsal digital
branches of
radial n.

6.12.4
Course of radial nerve; sensory supply to hand.

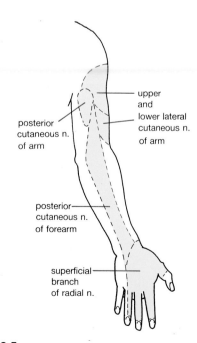

upper
and
lower lateral
cutaneous n.
of arm

posterior
cutaneous n.
of arm

posterior
cutaneous n.
of forearm

superficial
branch
of radial n.

6.12.5
Radial nerve: supply to skin.

cutting through the fibres of supinator; identify at least one branch to a forearm extensor muscle.

Now return to the main body of the radial nerve as it lies anterior to supinator and follow it distally through the anterior compartment of the forearm; you will find it lying between brachioradialis and the deeper muscles of the forearm (in particular, flexor pollicis longus). As it approaches the wrist, the nerve passes dorsally deep to the tendon of brachioradialis and then breaks up into superficial **dorsal digital branches** which supply the skin of the lateral $3\frac{1}{2}$ (or $2\frac{1}{2}$) digits on the back of the hand as far as the distal phalanx. It is possible to feel some of these branches by extending your thumb and palpating over the tendon of extensor pollicis longus.

The radial nerve is susceptible to injury in the upper arm. If long crutches are used incorrectly the body weight may be taken by the upper bar of the crutch pressing hard into the armpit. Usually, however, crutches are designed so that the body weight is supported by the grip of the hand on a lower bar or by supporting bars for the forearm. Excessive pressure into the axilla can produce a paralysis of the radial nerve and therefore of the extensor muscles of the forearm, with subsequent 'wrist drop'.

Qu. 12B   *If a patient had sustained a fracture of the shaft of a humerus and you suspected damage to the radial nerve, what would you expect to find on examination?*

B. Rapid assessment of the major nerves of the upper limb

When confronted with a patient who has a badly injured upper limb it is important to assess the function of the major nerves. However, it is often impossible, or too painful, for the patient to move the forearm, wrist, and fingers, so that reliance must be placed in the first instance on sensory function. A pinprick into the pulp of the terminal phalanx of the index finger will test the presence of median nerve function, and a similar pinprick of the little finger will test ulnar nerve function. Pinprick over the dorsum of the first intermetacarpal space will test the radial nerve, but is not quite so reliable. If extension of the thumb can take place at the interphalangeal joint then the radial nerve must be intact to the level of the posterior interosseous nerve.

Requirements:

Articulated skeleton
Prosections of the axilla, arm, forearm, and hand showing the posterior cord and its branches to the posterior wall of the axilla, the axillary nerve, and the course and distribution of the radial nerve.

Seminar 13

The breast

A. Development and function of the mammary gland

The mammary glands, like sweat and sebaceous glands, develop from the epidermal layer of the skin. Six or so columns of epidermal cells proliferate and invaginate the mesoderm along two lines (mammary lines) which extend from the axilla, sweep medially to the mid-clavicular line, and pass down to the groin. Normally, all but the pectoral pair of these invaginations (overlying the 3rd–5th ribs) disappear. Accessory breasts, or more commonly accessory nipples (the 'devil's marks' of medieval witch hunters), may be found at any point along the mammary lines but are more common in the pectoral region.

The pectoral pair of mammary primordia develop to form 12–15 main lactiferous ducts with a few branches. The lactiferous ducts open on the surface of a raised nipple which is surrounded by an area of pigmented skin, the areola. In the immature female and the male, the mammary glands are similar: the mammary tissue is rudimentary and confined to the margins of the areola. It comprises little more than a relatively unbranched ductal system in a connective tissue stroma. The areolae are pink and the nipples relatively undeveloped. The structure of the gland remains rudimentary until the time of puberty when, in the female, concentrations of plasma oestrogen increase and fat is laid down around the growing ductal system; the alveoli remain primitive. If pregnancy occurs, increasing plasma concentrations of oestrogen and progesterone are responsible respectively for proliferation of the ductal system and the development of alveoli around the terminal ducts; the areola often becomes more deeply pigmented. At parturition, production of milk is stimulated primarily by the secretion of prolactin from the anterior pituitary gland. Ejection of milk from the breast is caused by a reflex in which the neural stimulation of suckling leads to an output of oxytocin from the posterior pituitary. This hormone causes the contraction of a network of specialized myoepithelial cells ('basket cells') which surround the alveoli, thus expressing milk into the ducts. The suckling stimulus also maintains the secretion of prolactin. The term lactation implies both milk production and milk ejection.

After the menopause mammary tissue atrophies. The breasts tend to sag because the connective tissue which supports them earlier in life ('suspensory ligaments'), like the mammary tissue, is dependent on oestrogen from the ovaries. Oestro-gen replacement therapy can be used to reduce this change.

Although mammary tissue in the male and immature female is rudimentary, it can respond to hormones. Likewise, maternal hormones can cause the production of mammary secretions ('witches' milk') in newly born infants. In adult males, breast development may occur as a result of hormones produced abnormally in the body or administered for therapeutic purposes.

Examination of the living breast (6.13.1)

Each mammary gland lies in the superficial fascia largely over pectoralis major. Its form and size are very variable, but its base is more constant, extending from the level of the 2nd to 6th ribs between the sternum and the anterior axillary line. The base of a normal breast is mobile on the underlying chest wall. Any tethering to the deep structures can signify disease. An **axillary tail** extends upwards and laterally into the axilla. Some deep pockets of breast tissue may pierce the deep fascia. In the male, the **nipple** usually lies in the fourth intercostal space and the mid-clavicular line; in adult females its position is much more variable.

When examining the breasts their contours should be studied in a variety of positions of the arm to detect any irregularity or asymmetry. As the arm is moved, the breasts should move freely on the chest wall (see **6.13.5**) and the overlying skin should not be tethered. The nipples are usually everted. The pigmentation of the **areola** varies: **6.13.1** shows a normal, non-lactating breast and should be compared with the lactating breast in **6.13.2**.

Palpation of the breast is done initially with the palm of the hand rather than with the fingertips, which sense too many small normal fatty nodules in the breast tissue. If a lump or other abnormality is discovered, its attachment to other tissues should be determined by appropriate positioning of the patient or achieving contraction of pectoral muscles. The axillary lymph nodes should be examined for evidence of enlargement or tethering to surrounding tissues.

B. Radiology

A radiograph of a breast is referred to as a mammogram, and these are routinely used for screening in some breast clinics. Mammogram **6.13.3** shows a breast carcinoma (arrow) which distorts the local architecture and contains characteristic microcalcification.

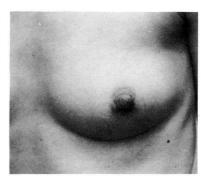

6.13.1
Non-lactating breast.

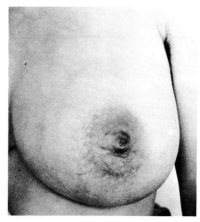

6.13.2
Lactating breast.

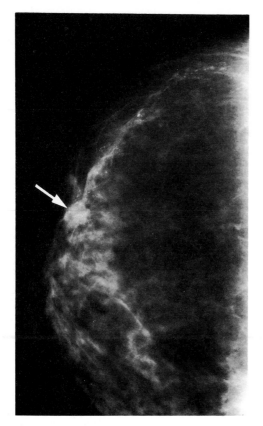

6.13.3
Mammogram: note presence of carcinoma with microcalcification (arrow).

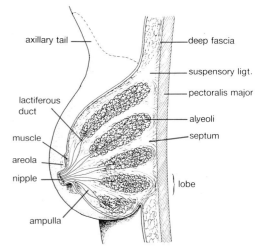

6.13.4
Components of breast and nipple.

C. Prosections (6.13.4)

Examine a dissected breast and identify some of the 15–20 **lobules** of glandular tissue arranged radially around the nipple and separated from each other by fibrous tissue septa. Each lobule is drained separately on to the nipple by its **lactiferous duct**, which is dilated to form an **ampulla** just before it opens on to the skin. The nipple is surrounded by circular muscle fibres which erect it during suckling.

Areolar glands open around the nipple. They secrete an oily sebaceous lubricant and are most prominent in pregnancy and lactation.

The breast is supported by fibrous **suspensory ligaments** which extend from the deep fascia overlying pectoralis major to the dermis. Other fibres pass to the skin from the connective tissue of the breast. If breast tissue is abnormally swollen with tissue fluid, these fibres invaginate the skin ('peau d'orange' appearance). In older women the fibrous supports atrophy and the breasts tend to become pendulous.

The **innervation** of the breast is from the segmental intercostal nerves (see Vol. 2). The **arterial** supply and venous drainage come from two main sources: the internal thoracic artery and intercostal arteries medially; and branches of the axillary artery laterally. These vessels enlarge considerably during lactation. The veins corres-

pond to the arteries. It is noteworthy that the intercostal veins communicate directly with the vertebral veins and that secondary malignant deposits in the vertebrae are common.

The **lymphatic drainage** of the breast (**6.13.5**) is extensive and is of considerable significance if the tissue undergoes malignant change. Lymphatic vessels around the ducts drain the glandular substance of the breast. Vessels running between the lobules and lobes drain to a **subareolar plexus** and to a **submammary plexus** in the deep fascia of the muscles covering the chest wall. From these plexuses lymph passes centrifugally to the regional nodes.

The majority of lymph from the breast drains to the **axillary nodes** but some drains to other nodes, particularly the **internal thoracic** group within the chest. The axillary nodes can be subdivided into a number of overlapping groups: a pectoral group behind the lateral border of the pectoralis major; a lateral (brachial) group along the course of axillary vein; a subscapular group along the course of the subscapular artery; a central group on the floor of the axilla; and an apical group in the apex of the axilla. The apical group drains the other groups and gives rise to a single **subclavian lymph trunk** which passes upward with the subclavian vein to empty into its junction with the internal jugular vein or, on the left side, into the thoracic duct.

These various groups of nodes intercommunicate but, in general, the pectoral group of nodes receive most of the breast lymph, passing it directly or indirectly to the apical group. The lateral group receive most of the lymph from the arm, and the subscapular group from the back of the upper part of the trunk and axillary tail of the breast, if it lies deep to the deep fascia. The internal thoracic group of nodes drain the medial part of the breast (there is a node in most of the upper six intercostal spaces alongside the internal thoracic artery). The internal thoracic nodes are important because they are often involved at an early stage in the spread of breast cancer; moreover they communicate freely across

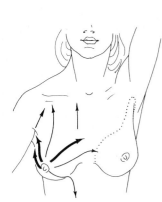

6.13.5
Direction of lymphatic drainage from breast.

the midline behind the sternum with corresponding glands of the opposite side. Spread of cancer cells from these nodes can lead to escape of cancer cells into the pleural cavity, creating secondary deposits, effusion of fluid into the cavity, and rapid progress of the disease.

One or two nodes may lie in the delto-pectoral groove along the cephalic vein, and others in the infraclavicular fossa. They receive some lymph from the upper part of the breast, and drain through the clavipectoral fascia (which lies between the clavicle and subclavius above, and pectoralis minor below) to reach the apical axillary nodes. Lymph from the lower part of the breast can drain to vessels and nodes in the upper abdominal wall, and cancer cells can spread via this route to the peritoneal cavity and liver.

Surgical treatment of breast cancer has varied considerably over the years. A radical ablation procedure involves removal of all breast tissue, the underlying muscles, and as many as possible of the nodes. Current surgical practice is usually to remove the local lesion and to combat spread of the disease either by radiotherapy directed at the various groups of potentially affected nodes, or by chemotherapy.

Requirements:

Prosections of the female mammary gland showing the lobules of secretory tissue and their ducts opening onto the nipples

Mammograms and xeroradiographs of the breast.

CHAPTER 7

The lower limb: introduction

Bipedalism in hominids means that the lower limb has become modified from the general pentadactyl pattern to function in both stance and the many varied forms of locomotion that are undertaken, often on very irregular surfaces. Unlike the arm, the lower limb frequently bears the entire weight of the body and this is regularly taken on just one leg in walking. Moreover, the stresses that the limb has to withstand are often multiplied many times, as on landing after a jump. In adapting to these functions the lower limb has forgone specializations which enable the upper limb to perform precision movements over a wide range. Nevertheless, considerable dexterity can be attained with practice as, for example, in a person unable to use the upper limbs for this purpose.

The weight-bearing function of the lower limb demands that the joints are modified for stability in weight-bearing. The weight of the body is transmitted through the shaft of each femur largely to the tibia in the lower leg. This bone, with the fibula, forms a mortice-type joint transmitting weight on to the talus, the bone which forms the apex of the arch of the foot. Finally, the arch transfers weight forward to the toes and backward to the heel.

The lower limb also requires considerable mobility in locomotion. To achieve this, the neck of the thigh bone has become elongated so that the shaft is offset from the head, an arrangement which also gives greater leverage to muscles acting at the upper end of the femur. Since the presence of the neck increases the distance between the upper ends of the femoral shafts, which are separated more widely than are the feet at ground level, the shafts slope downward and medially.

Furthermore, as a result of the medial rotation of the lower limbs during development not only are they brought closer together but the powerful extensor muscles of the thigh, leg, and foot are positioned anteriorly, with the big toe lying medially. This is in contradistinction to the upper limb with its manipulative functions where the flexor muscles face anteriorly and the homologue of the big toe, the thumb, lies laterally. A major advantage which results from the different arrangement of muscles in the lower limb is propulsion where the combined strength of the extensor muscles can be exerted on the bones and joints to propel the body forward.

While studying the lower limb you should carefully consider the anatomical specializations concerned with stance, which consists of weight-bearing on one or both legs; with walking, which consists of a stance phase and a swing phase for each limb in turn; and with other forms of gait such as running and jumping (see Ch. 9).

Seminar 1

Bones of the lower limb

Aims To study the bones of the pelvic girdle and lower limb, their surface contours and internal architecture which reflect the various forces and articulations to which they are subject; to study their growth and development. Whilst doing so you should also consider their skeletal functions and compare their functional adaptation with that of the upper limb bones.

A. Living anatomy and the bony skeleton

As you study the bones of the lower limb, consider the general functions of the bony skeleton, the adaptations of the lower limb to the upright stance and bipedal locomotion (see p. 153), and the developmental rotation of the lower limb and its consequences (p. 29).

Identify on a skeleton and on the radiographs the bony points listed below (**7.1.1, 7.1.2, 7.1.3**). Note which of the bony prominences can be felt in yourself or your partner.

Pelvic girdle

The bony pelvis has several important functions, one of which is the distribution of the weight of the head and neck, upper limbs, and torso to the lower limbs.

The pelvic girdle comprises two **innominate bones**, each made up of a fused **ilium, ischium**, and **pubis**, joined together in the anterior mid-line by the **pubic symphysis**, and joined to the sacrum by synovial sacroiliac joints. Movements at the sacroiliac joints are almost entirely prevented by strong ligaments and interlocking articular cartilage contours. Identify on the bony pelvis and yourself, where this is possible:

Pubic bone

- body of the pubis
- pubic crest
- pubic tubercle
- superior and inferior pubic ramus
- pectineal line (sharp posterior border of the upper aspect of the superior pubic ramus) which extends laterally into the iliopectineal eminence.

Ischium

- ischial tuberosity
- ischial spine
- lesser sciatic notch (between tuberosity and spine)

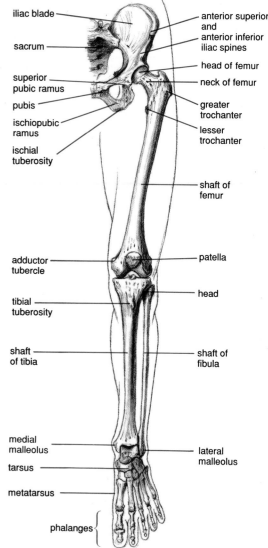

7.1.1
Bones of lower limb and pelvic girdle; anterior view.

- greater sciatic notch (above ischial spine)
- ischiopubic ramus (fused ramus of the ischium and inferior ramus of the pubis)

Ilium

- blade of the ilium, hollowed on its inner aspect to form the iliac fossa
- iliac crest and its tubercle

- anterior superior and anterior inferior iliac spines
- posterior superior and posterior inferior iliac spines
- articular surface for sacrum
- anterior, inferior, and posterior gluteal lines on outer surface of iliac blade (see **7.3.2**)

The **acetabulum**, with its horseshoe-shaped articular surface and acetabular notch, is formed from the three pelvic bones.

Three **buttresses** of bone resist weight-bearing stresses:
- when standing on both legs, weight is distributed from the sacral articular facet to the acetabulum;
- when supported on one leg, the line of maximum stress is directed vertically upward from the acetabulum to the tubercle of the iliac crest;
- when seated, weight is distributed from the sacral articular facet to the ischial tuberosity.

Femur

- head with a small central pit
- neck—note the angle it makes with the shaft
- greater trochanter overhanging the trochanteric fossa
- lesser trochanter
- anterior intertrochanteric line and posterior intertrochanteric crest
- gluteal tuberosity, a short rough ridge extending downward from the base of the greater trochanter to the linea aspera
- shaft with posterior linea aspera (rough line) which divides into the medial and lateral supracondylar lines at its lower end
- popliteal surface
- medial and lateral condyles
- medial and lateral epicondyles
- intercondylar fossa
- groove for the tendon of popliteus on the lateral condyle
- adductor tubercle

Qu. 1A *What is the functional significance of the neck of the femur? In what way is the neck of the femur comparable with the clavicle?*

Patella

- Posterior articular surfaces for the condyles of the femur; their lower limit indicates the line of the knee joint cavity

Tibia

- upper surface ('plateau')
- medial and lateral condyles
- tibial tuberosity

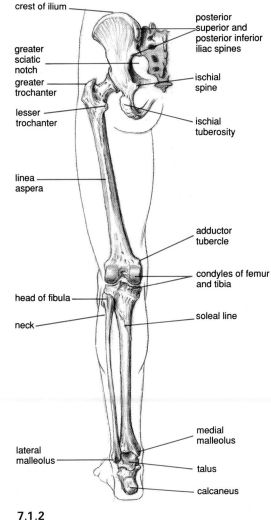

crest of ilium

greater sciatic notch

greater trochanter

lesser trochanter

linea aspera

head of fibula

neck

lateral malleolus

posterior superior and posterior inferior iliac spines

ischial spine

ischial tuberosity

adductor tubercle

condyles of femur and tibia

soleal line

medial malleolus

talus

calcaneus

7.1.2
Bones of lower limb and pelvic girdle; posterior view.

C = calcaneus
T = talus
N = navicular
Cu = cuboid
M = medial cuneiform
I = intermediate cuneiform
L = lateral cuneiform

7.1.3
Tarsal bones; dorsal view.

- shaft—with anterior (subcutaneous) and interosseous borders; and medial, posterior, and lateral surfaces

- soleal line

- laterally-placed facets for articulation with the head and lower end of the fibula

- medial malleolus—with laterally-facing facet for articulation with the talus

Qu. 1B *What is the functional significance of the flat upper surface of the tibia?*

Fibula

- head—with apex and facet for articulation with the tibia

- neck

- shaft—with interosseous border; and the lateral, posterior, and medial surfaces

- inferior end—with surface for interosseous ligament and facet for articulation with talus

- lateral malleolus and malleolar fossa

Tarsus

- **talus**—with facets for articulation with tibia and fibula, and with calcaneus and navicular

- **calcaneus**—with facets for articulation with talus and cuboid; posteriorly placed medial and lateral tubercles on its undersurface; large medial protuberance, the sustentaculum tali

- **navicular** and its tuberosity

- medial, intermediate, and lateral **cuneiform** bones

- **cuboid** and groove for peroneus longus tendon

Metatarsus

Each of the 5 metatarsals has a

- base

- shaft

- head

The first metatarsal is shortest and strongest. Two sesamoid bones (in flexor hallucis brevis) at its distal end are not usually retained in articulated skeletons. The base of the 2nd metatarsal is long and is held firmly in a mortise between the cuneiform bones. This is thought to limit the movement of this metatarsal, and to account for the frequency with which it undergoes stress fracture on sustained walking ('march fracture')

Phalanges, proximal, middle, and distal

- base

- shaft

- head

The great toe, like the thumb has only two phalanges, which are very strong. In comparison with those of the hand, the middle and especially the distal phalanges of the toes are much reduced.

The weight of the body is distributed between the posterior tubercles of the calcaneus and the heads and sesamoid bones of the metatarsals, the intervening bones forming an arch.

Qu. 1C *What is the functional significance of the arch of the foot?*

Having identified the bony prominences, you should now consider the position of the lower limb. Viewed anteroposteriorly, the hip, knee, and ankle joints should lie one above the other (**7.1.2** and see **9.1**); viewed laterally, the femur and tibia both slope forward at 5° from the vertical so that the hip joint lies anterior to the knee and ankle joints (see **9.1**).

Assessment of leg length

It is often necessary to measure the length of a patient's lower limbs and, if a discrepancy between them is found, to determine the source of the difference.

With your partner lying supine on the couch, abduct one leg slightly, and then adduct the opposite leg, so that the two limbs again lie parallel to each other, though at an angle to the trunk. You will see that the abducted leg looks longer than the adducted leg. A measure of the difference between the 'apparent' lengths of the legs in this position can be made by firstly reading off the distance on a tape measure between the umbilicus and the medial malleolus of the abducted leg, and subtracting from it the similar measurement of the adducted leg; or simply by measuring the distance between the two malleoli. Do this and record your findings. Such 'apparent' shortening or lengthening of a leg may occur in a patient whose pelvis is tilted because of a fixed curvature of the spine, or because of an abduction or adduction deformity of one of the hip joints. The difference in leg length is called 'apparent' because it is not due to any 'real' difference in the lengths of the legs, but to the position in which the legs lie in relation to the trunk.

The 'real' length of the leg can be measured from a fixed point on the pelvis. The anterior superior iliac spine is a convenient point. Record the measurement of your partner's legs from the anterior superior iliac spine to the medial malleolus on each side.

Are the lengths the same? Suppose there is an inch difference, how would you tell where the discrepancy occurs? (it could, for instance, be in the leg below the knee).

Ask your supine partner to bend his knees to a right angle. If one femur is shorter than the other the difference becomes obvious and can be measured. Similarly if one shin is shorter than the other the difference is obvious, and can be confirmed by a measurement from the joint line of the knee to the medial malleolus.

The tip of the greater trochanter normally lies at the same horizontal level as the centre of the head of the femur. Confirm this on a skeleton, and feel your partner's greater trochanters. You can judge if they lie at the same level, or you can measure the distance between a fingertip placed on the greater trochanter and the highest point of the iliac crest. If there is a discrepancy between the level of the greater trochanter on the right and left, then the true difference in leg length lies in the region of the neck of the femur or hip joint. Perhaps the angle of the neck of the femur to the shaft is much reduced, or the joint is dislocated, or the neck of the femur is fractured! The length of the shaft of a femur can be measured from the tip of the greater trochanter to the line of the knee joint. The length of a femur is said to be about one quarter of the height of a body.

Record all the measurements you make.

B. Radiology and development of bones of lower limb

Review pp. 36–7 which are as relevant to the lower limb as to the upper limb.

Qu. 1D *Examine 7.1.4, 7.1.5, 7.1.6 and comment on the obvious trabecular patterns. 7.1.6 is of an aged oriental woman whose feet were 'bound' from birth.*

Examine **7.1.7** which is a radiograph of the pelvis and upper end of the femur of an adult. Compare it with **7.1.8** which is of a newborn baby in which the innominate bone is made up of three separate centres, the ilium, the pubis, and the ischium. At this young age the acetabulum is entirely cartilaginous, at the meeting place of its separate elements, and the centre for the head of the femur is not yet visible. **7.1.9** is a radiograph of the hips of an infant

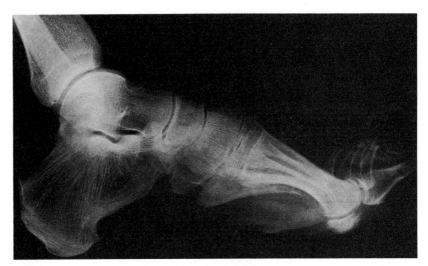

7.1.5
Trabeculae of bones of foot.

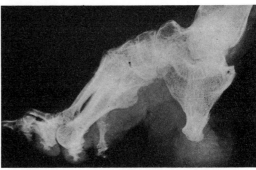

7.1.6
Bones of an oriental foot distorted by binding since birth.

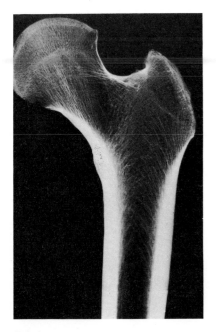

7.1.4
Trabeculae of upper end of femur.

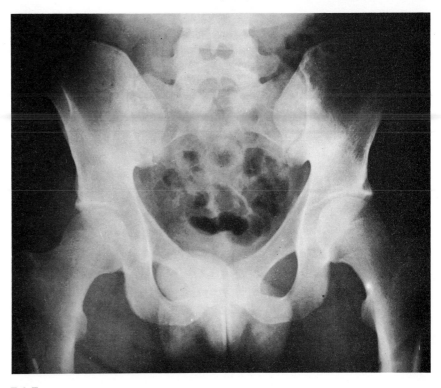

7.1.7
Pelvis and hip joints of adult female.

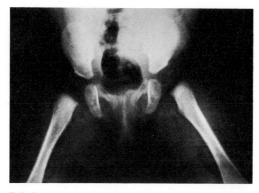

7.1.8
Pelvis and hip joints of neonate.

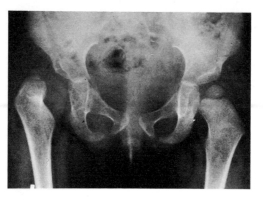

7.1.9
Pelvis and hip joints of child aged 6 months.

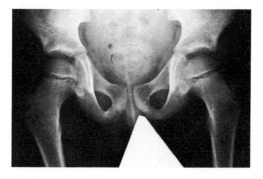

7.1.10
Pelvis and hip joints of child aged 7 yrs.

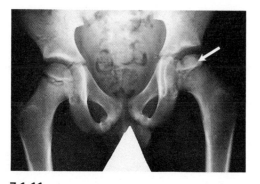

7.1.11
Disordered capital epiphysis of the left femur (arrowed) in child aged 8 years.

six months old; the centre of ossification of the head of the femur is now visible.

i) On the right side of **7.1.9** draw a line which passes across the pelvis horizontally from one acetabulum to the other, through the points where the separate elements of the innominate bone meet. Next draw a vertical line from the outer edge of the ilium, where it forms the outer lip of the acetabulum, to intersect the horizontal line. The centre of ossification of the head of the femur should lie in the lower inner quadrant.

ii) On the right side of **7.1.9** draw a curved line connecting the inferior aspect of the neck of the femur with the inferior margin of the superior pubic ramus. This line, known as **Shenton's line**, is smooth and continuous (see also p. 100, Qu. 2B).

7.1.10 is a radiograph of the pelvis of a child aged 7 years. The centre of ossification of the head of the femur is now much larger, and forms a quite large 'cap' over the upper end of the femur, in the acetabulum. The acetabulum is more fully formed, and the centre of ossification of the greater trochanter is appearing. The blood supply to the epiphysis of the head of the femur, which comes almost entirely from vessels of the trochanteric anastomosis (pp. 100, 127) which run upward on the posterior aspect of the neck of the femur, is vulnerable between the ages of five and ten years. If this blood supply is impaired, the capital epiphysis dies in whole or in part, and the bone which later reforms in the head of the femur may become flattened and misshapen. **7.1.11** shows the result of such a condition, known as Perthes' disease. Compare the appearance of the diseased hip with the normal, opposite, hip.

In a 15-year-old boy the head of the femur is well developed (**7.1.12**), and an obvious growth plate can be seen between the head and neck of the femur. During adolescence, great forces are transmitted through this cartilaginous growth plate and it is not surprising that, if the cartilage structure is not entirely normal, the head of the femur can 'slip' on the neck, either acutely or more usually as a subacute process. The head of the femur slips inferiorly (seen on an A–P radiograph) and posteriorly (seen on a lateral radiograph), so that the leg tends to take up a position of some lateral (external) rotation, with some limitation of abduction at the hip joint. This condition is known as

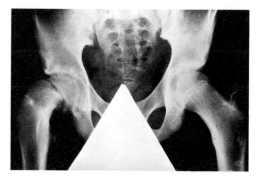

7.1.12
See Qu. 1E

'slipped epiphysis', and is unique to the hip joint. Draw a line along the upper margin of the neck of the femur in the normal hip joint and notice that it passes through the upper third of the capital epiphysis. This is one way of recognizing an early stage of slipping of the epiphysis in a young person complaining of pain in the hip, thigh, or knee.

Qu. 1E *Draw this line on both sides of 7.1.12. Which is the normal side?*

Examine **7.1.13**, a radiograph of a young patient (with plaster casts on both feet). Note the presence of epiphysial plates at the lower end of the femur and upper ends of the tibia and fibula. Note also that the epiphysial plate of the lower end of the fibula lies at a lower level than that of the tibia and at the same level as the line of the ankle joint.

For radiographs of the bones comprising the adult knee and ankle see **7.2.9**, **7.2.10**, **7.2.24**, and **7.2.25**.

Examine **7.1.14a,b** and **7.1.15a,b** and **7.1.16a,b** which show the feet of children respectively aged eighteen months, four years, and eleven years, noting the development of the ossification centres. For a radiograph of the adult foot, see **7.1.5**.

Requirements:

Articulated skeleton
Separate bones of the pelvis, leg, and foot
Radiographs of the adult and juvenile bones of the lower limb.

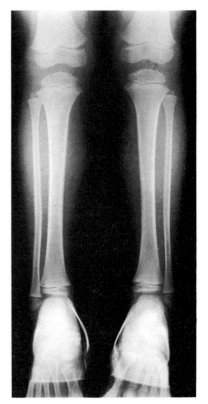

7.1.13
Bones of leg of child aged 6 yrs.

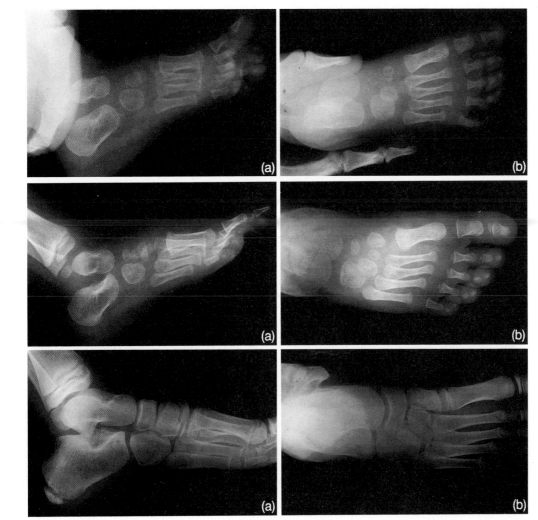

7.1.14
Foot of child aged 18 months: (a) lateral view; (b) dorsal view. (Note the adult hand holding the foot.)

7.1.15
Foot of child aged 5 years: (a) lateral view; (b) dorsal view.

7.1.16
Foot of child aged 11 years: (a) lateral view; (b) dorsal view.

Seminar 2

Joints of the lower limb

Aims To study the movements which occur at the principal joints of the lower limb; to study the structure of these joints, their classification, and their structural adaptations for movement and stability. As you study the joints you should consider their adaptation for the upright stance and bipedal locomotion (Ch. 9), and compare their functional adaptation with that of the joints of the upper limb.

THE HIP JOINT

A. Living anatomy

Your trunk is essentially balanced on a transverse axis passing through the hips. In normal stance the line of gravity of the trunk passes through or just behind the hip joint, creating a slight extensor moment (p. 153).

Look again at the articular surfaces on the skeleton and the radiographs. Move your own hip joint and note the type and range of movements which you can perform. The movements are similar in type to those at the glenohumeral joint because both are ball and socket synovial joints. They are, however, more restricted in their range. Furthermore, whereas shoulder movements involve both the glenohumeral joint and scapular mobility, the individual pelvic bones are essentially immobile.

Hold your partner's right foot and ankle firmly and, with his leg fully extended, rotate the leg both medially and laterally. These movements are taking place at the hip. Test the degree of abduction and adduction which you find at the hip and record your findings.

Get your partner to lie on his side on the couch and pull his straightened leg backwards. Such extension as takes place is due largely to movement of the pelvis rather than the hip. Now flex your partner's hip until the thigh rests against the trunk. The normal hip joint appears to flex about 130° but, in fact, the hip joint itself can usually only flex about 90°–100°. Now, with your partner lying on his back, put your hand behind the lumbar spine and again flex the hip joint. As the hip passes a right angle you will notice a flattening of the normal gentle forward curve of the lumbar spine (lordosis). Any flexion of the hip which occurs thereafter is caused by rotation of the pelvis around a transverse axis. Sometimes, in disease of a hip joint, the capsule contracts from scar tissue so that a flexion deformity develops and the patient cannot extend the hip joint from the flexed to the anatomical position. However, such a flexion contracture is disguised if the lumbar lordosis increases so that the

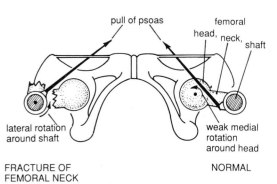

FRACTURE OF   NORMAL
FEMORAL NECK

7.2.2
Diagram (viewed from feet) to show line of pull of iliopsoas, with neck of femur either intact or fractured.

pelvis is rotated sufficiently to permit the thigh to reach the anatomical position with respect to the trunk (**7.2.1**).

An understanding of how muscles produce medial and lateral rotation of the hip joint is complicated by the presence of the femoral neck. The neck offsets the shaft and greater trochanter (to which the muscles are attached) from the axis of hip rotation which, in the standing position, passes vertically down through the centre of the femoral head to its lateral condyle. Medial and lateral rotation occur with each step, as the pelvis swings. In the upright posture, muscles which pass in front of the axis produce medial rotation, even if, like adductor longus, they are attached to the posterior surface of the femur; muscles which pass behind the axis produce lateral rotation.

If, however, the neck of the femur is broken, the shaft of the femur is free to rotate about its own longitudinal axis without constraint from the hip joint. Under these conditions, powerful muscles, such as psoas major, will produce lateral rotation of the femoral shaft. Patients with a fractured neck of femur therefore classically present with the uninjured leg in the normal position, and the injured leg in marked lateral rotation (**7.2.2**, **7.2.3**, **7.2.4**).

B. Prosections

Examine the capsule of the joint (**7.2.5**, **7.2.6**) and note its attachment to the margins of the acetabulum and to the transverse ligament which bridges the acetabular notch. The distal attachment of the capsule is to the intertrochanteric line of the femur anteriorly, but posteriorly the capsule is attached around the middle of the neck of the femur. The capsule is reinforced by an anterior **ilio-femoral ligament**; the ligament is shaped like an inverted

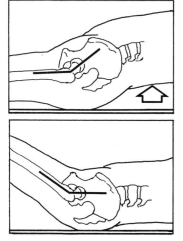

7.2.1
Test to detect fixed flexion deformity of hip joint.

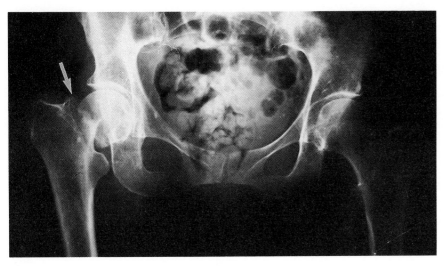

7.2.3
Radiograph of fractured neck of right femur
(arrowed).

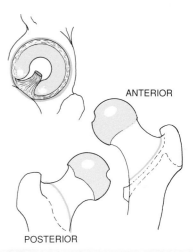

7.2.4
Position of leg in a recumbent
patient after fracture of neck of
right femur.

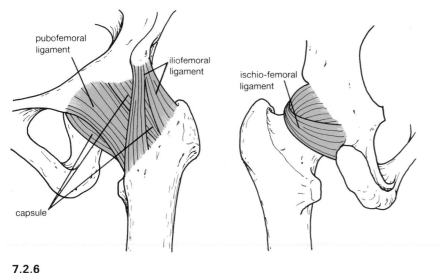

7.2.6
Intrinsic ligaments of hip joint.

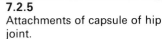

7.2.5
Attachments of capsule of hip
joint.

Y with its stem attached to the anterior inferior iliac spine, and its two limbs attached to the upper and lower ends of the intertrochanteric line of the femur. The ilio-femoral ligament is extremely strong and can act as a fulcrum when a dislocated head of femur is being repositioned (reduced) in the acetabulum. The capsule is also reinforced by the **pubo-femoral ligament** which runs from the ilio-pubic eminence to the inferior aspect of the neck of the femur. Although an **ischio-femoral ligament** is described as arising from the ischium posterior to the acetabulum, many of its fibres have no distinct attachment to the femur, but merge in a circular fashion into the capsule. The major parts of all three ligaments therefore pass spirally from their origin on the hip bone to limit extension.

Examine the inside of the joint, noting the horseshoe-shaped **articular cartilage** of the acetabulum, the **pad of fat** which occupies the rest of the acetabular surface, and the **round ligament of the head of the femur** (ligamentum teres). The round ligament arises from the transverse ligament of the acetabulum and attaches to a pit in the head of the femur (the blood vessel it carries supplies only the small area of the head to which the ligament is attached). As in the shoulder joint the depth of the acetabulum is increased by a ring of fibrocartilage around its margin (**acetabular labrum**). Synovial membrane lines the capsule and non-articular structures within the joint such as the pad of fat and the round ligament.

On the neck of the femur within the distal part of the capsule you may see, beneath the synovial membrane, some fibrous bands (**retinacula**) which carry blood vessels to the foramina of the neck and head of the femur (**7.2.7**). If these vessels are damaged owing to a fracture of the neck, which is not uncommon in elderly people, the head may die and then crumble, destroying the joint.

Qu. 2A   *With respect to stability and movement what are the major differences between the synovial 'ball and socket' joints of the upper and lower limbs?*

Look again at **7.1.7**, the radiograph of the pelvis and hip joints of an adult. Now re-examine **7.1.9**, a radiograph of the pelvis of an infant six months old. About one in a thousand babies are born with one or both hip joints liable to develop in a dislocated condition so that it is important to be able to determine whether or not the head of the femur is in the acetabulum at this time.

Qu. 2B   *Draw the line described in (i) on p. 96 on both sides of 7.1.9. Which hip is dislocated?*

Qu. 2C   *Why is there a lower incidence of 'congenital dislocation' of the femur in Nigeria, where many infants are carried with their legs astride on their mother's hip?*

Examine the vertical MRI of the normal pelvis and hip joint (**7.2.8**), noting the articulating surfaces of the head of the femur and acetabulum. MRI studies of the hip joint are particularly useful in conditions where a diminished blood supply to the femoral head leading to ischaemic necrosis of the bone is suspected. As you study the muscles around the hip joint, try to identify them on this image.

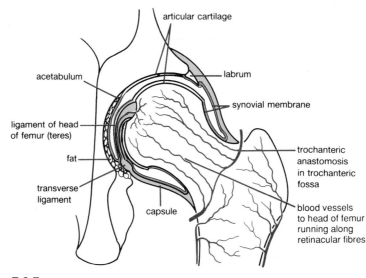

7.2.7
Arterial supply to head of femur.

7.2.8
MRI of pelvis and hip (coronal section). H—head of femur; Ca—calcar femorale in head and neck of femur; N—neck of femur; P—body of pubis; I—ilium; GMe—gluteus medius; GMi—gluteus minimus; OE—obturator externus; OI—obturator internus; Ad—adductor muscles; Ps—psoas; Il—iliacus; B—bladder; Pr—prostate.

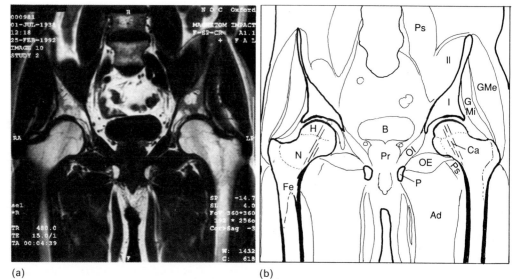

(a)   (b)

THE KNEE JOINT

A. Living anatomy

Get your partner to stand upright. Both patellae should face forward and the knee joint should lie immediately below the hip joint. In small children up to the age of four years the knees are often in contact when the feet are up to 10 cm apart ('knock knees'). This 'valgus' deformity usually disappears with time.

Look again at the articular surfaces of the knee joint on the skeleton and examine the radiographs **7.2.9** and **7.2.10**. Note that the fibula takes no part in the knee joint.

Move your own knee joint. It can be flexed and extended and therefore appears to be a simple synovial hinge joint. However, other movements are possible, so it is classified as a compound hinge joint. With your knee flexed at a right angle try to rotate your tibia medially and laterally; about 40° of movement is possible. Now extend your knee fully and again try to rotate your tibia; no movement occurs at the knee joint. If a flexed knee joint is violently rotated while weight-bearing (as when kicking a football) one of the intra-articular cartilages (menisci) may be split.

More subtle movements take place at the knee which cannot easily be observed in the living. Stand with your knees very slightly flexed and then extend them fully. You may be able to feel that, as full extension is approached, the femur rotates medially on the tibia by a few degrees. This is the process of 'locking' of the knee joint. In the locked position all the ligaments are taut and virtually no muscular effort is required to support the body on a locked knee. However, before the fully extended knee can be flexed, it must be actively 'unlocked' by a muscle (popliteus; see below) which rotates the femur laterally on the tibia.

With your partner seated, try to move the upper end of the tibia forward and backward on the femur; no movement should occur (p. 103; **7.2.17**).

Define, on your own and your partner's knee joint the following parts of the joint:

- the full circumference of the patella, especially its medial border
- the articular margin of each femoral condyle
- the articular margin of the anterior parts of each tibial condyle
- the joint line anteriorly, medially, and laterally
- the medial and lateral epicondyles of the femur
- the tibial tubercle
- the adductor tubercle.

B. Prosection

Examine a prosected knee joint (**7.2.11, 7.2.12**) and note the extent of the hyaline articular cartilage which covers the condyles of the femur, the lateral and medial tibial condyles, and the posterior aspect of the patella. Note also the kidney- and semilunar-shaped fibrocartilaginous **menisci** which lie between the condyles of the femur and tibia.

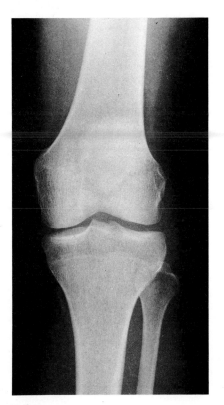

7.2.9
Knee joint (AP).

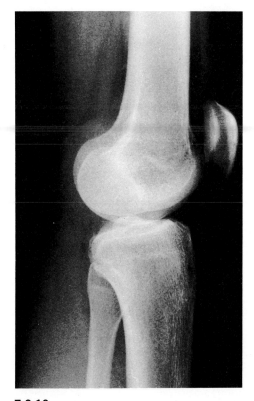

7.2.10
Knee joint (lateral).

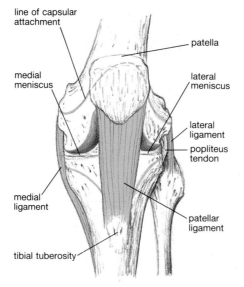

7.2.11
Knee joint: anterior view showing capsule and ligaments.

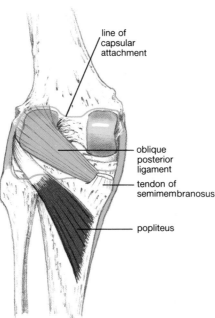

7.2.12
Knee joint: posterior view showing oblique posterior ligament and attachment of popliteus.

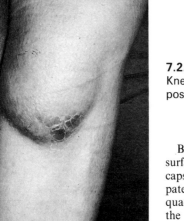

7.2.13
Swollen prepatellar bursa.

derived from quadriceps extend on to the tibial condyles. This arrangement protects the joint anteriorly. Posteriorly the capsule of the knee joint is attached to the femoral shaft above the condyles and to the margins of the tibial condyles. The **medial ligament** of the knee joint is broad and flat; it extends from the medial epicondyle of the femur to the antero-medial aspect of the tibia well below the medial tibial condyle; it is also attached to the medial meniscus. The **lateral ligament** is round and rope-like; it is attached superiorly to the lateral epicondyle of the femur and inferiorly to the head of the fibula. It is *not* attached to the lateral meniscus but is separated from it by the **tendon of popliteus**. This muscle, which is triangular in shape, arises from the posterior aspect of the tibia above an oblique line which runs across the upper part of the shaft. The tendon of popliteus passes upward through the capsule into the knee joint to attach to the lateral meniscus, and continues on to insert into a pit beneath the lateral epicondyle of the femur. The obliquely running **posterior ligament** spreads upwards and laterally from the posterior aspect of the medial tibial condyle where semimembranosus is inserted. It reinforces the capsule against torsional stresses and limits the rotation of the femur that occurs as the knee 'locks'. The **cruciate ligaments** are considered in the next section.

Intracapsular structures

The **synovial membrane** lines the capsule and those surfaces not covered by articular cartilage. Posteriorly, it covers the anterior but not the posterior surfaces of the cruciate ligaments and therefore it does not line the whole of the posterior aspect of the capsule. Anteriorly, synovial membrane lines the patellar ligament and the patellar retinacula; it also extends above the upper pole of the patella, as the so-called **suprapatellar bursa**. Therefore, when a knee joint swells because of bleeding within it, or because excess joint fluid has accumulated as the result of inflammation, the swelling is a diffuse one which extends proximally above the patella. This contrasts with the swelling that results from enlargement of the **prepatellar bursa**—'housemaid's knee'—which is confined to the front of the knee cap and does not communicate with the joint cavity (**7.2.13**). The **infrapatellar bursa** between the patellar ligament and the skin can also become inflamed and swollen ('clergyman's knee'). There are also several bursae behind the knee joint. It is not necessary to remember them all, but one situated between the medial head of gastrocnemius and the capsule of the knee joint can enlarge and form a cyst behind the knee.

Examine the **intra-articular pad of fat** which lies below the patella. It is covered with synovial membrane and attached by it to the intercondylar notch of the femur. The fat pad has two wing-like expansions known as the **alar folds**. This arrangement increases the surface area of the synovial membrane, helps to distribute the synovial fluid, and helps to fill the spaces which change in shape during movement of the joint.

Because the patella articulates with the anterior surface of the lower end of the femur, there is no capsule on the anterior aspect of the joint. The patella is a large sesamoid bone in the tendon of quadriceps which has become incorporated into the knee joint. Quadriceps inserts into the patella superiorly while inferiorly the **patellar ligament** attaches the patella, and thus quadriceps, to the tibial tuberosity. Fascial expansions (**retinacula**)

The fibro-cartilaginous **menisci** (7.2.14, 7.2.15) are attached to the irregular central part of the tibia by each of their two horns. Peripherally they are attached by **coronary ligaments** to the capsule from which they gain a blood supply. This is insufficient to nourish the inner parts of the cartilage which therefore rely on the synovial fluid. When damaged menisci are removed, their peripheral parts may subsequently regrow. The medial meniscus is partly fixed by its attachment to the medial ligament, whereas the lateral meniscus is not attached to the lateral ligament but to popliteus which controls its movements. Severe strain (usually rotational) on a meniscus can result in a longitudinal, or some times transverse, split in the fibrocartilage. This occurs more frequently in the medial meniscus because it is attached to the medial ligament of the knee joint and presumably not as mobile as the lateral meniscus (**7.2.14**). The detached piece of meniscus may move into the centre of the joint and prevent the knee from extending fully—the 'locked' knee of sporting injury (not to be confused with the 'locking' which occurs naturally on full extension). **7.2.15** is an arthrogram of the knee; air has been introduced into the joint to show up the meniscus.

The **cruciate ligaments** (**7.2.16**) are two thick rounded cords. The **anterior cruciate ligament** is attached to the anterior aspect of the upper surface of the tibia; it passes upward and backward to the inner aspect of the lateral condyle of the femur. The **posterior cruciate ligament** is attached to the posterior aspect of the upper surface of the tibia and passes upward and forward to the inner aspect of the medial condyle of the femur. These ligaments play an important part in the stability of the joint. If the anterior cruciate ligament is ruptured the tibia can be moved forward with respect to the femur (the anterior drawer test; **7.2.17a**); if the posterior cruciate ligament is ruptured, the leg will sag visibly if supported horizontally at the ankle (**7.2.17b**).

'Locking' and 'unlocking' of the knee joint

Take an isolated dissected knee joint and, with the tibia held firmly, gradually extend the femur. As full extension is approached, the femur will undergo a small degree of medial rotation on the tibia. The rotation occurs because, in the process of extension, the cruciate ligaments become taut sequentially and the medial condyle moves backward on the tibia. As the extension and medial rotation increase both the cruciate ligaments and the oblique posterior ligament become taut, further extension becomes impossible, and the joint 'locks'- (becomes close-packed). In addition, because of the cam shape of the condyles, the medial and lateral ligaments of the knee joint are also taut in full extension so that the joint is very stable. It is currently thought that, other than as a result of direct and violent trauma, the fully extended knee joint cannot be flexed unless popliteus contracts. This muscle causes the femur to rotate laterally on the tibia thereby 'unlocking' the joint.

Qu. 2D *If a footballer sustained an injury to a knee, and on examination it was found that the flexed lower leg could be moved backward and forward on the femur more easily than on the uninjured side, what would be the most likely cause?*

Qu. 2E *If another player complained of pain in the left knee, which he was unable to straighten, what structure might he have damaged?*

MR images are used in the evaluation of a wide spectrum of disorders of the knee. Examine **7.2.18** and **7.2.19**, noting in particular the cruciate ligaments and the tendon of popliteus. When you have studied the muscles and vessels of the lower limb identify these also on the images.

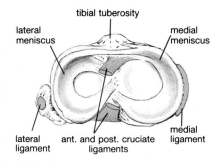

7.2.14
Menisci and cruciate ligaments of left knee viewed from above.

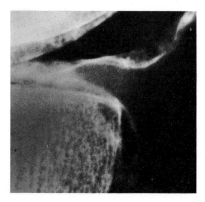

7.2.15
Arthrogram showing lateral meniscus.

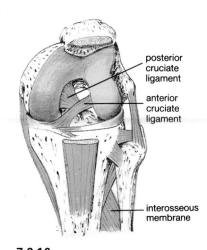

7.2.16
Cruciate ligaments, anterior view.

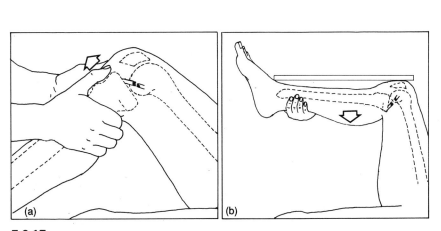

7.2.17
(a) Anterior 'drawer' test (see text). (b) 'Sag' test (see text).

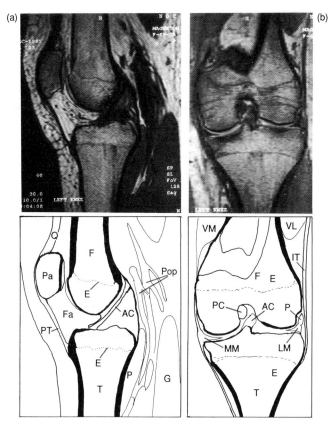

7.2.18

(a) Sagittal and (b) coronal MRIs of knee joint: Q—quadriceps tendon; P—patella; PT—patellar tendon; F—femur; T—tibia; E—epiphysial line in femur/tibia; Fa—fat pad in knee joint; AC—anterior cruciate ligament; PC—posterior cruciate ligament; Pop—popliteal vessels; P—popliteus; G—gastrocnemius; LM—lateral mensicus; MM—medial meniscus; IT—iliotibial tract; VM—vastus medialis; VL—vastus lateralis.

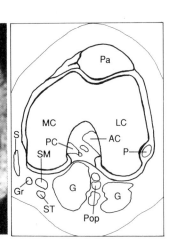
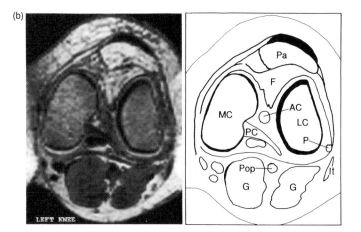

7.2.19

Transverse MRIs of knee joint (b) at lower plane than (a): Labels as 7.2.18 and MC—medial condyle; LC—lateral condyle; S—sartorius; Gr—gracilis; ST—semitendinosus; SM—semimembranosus.

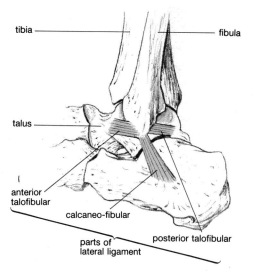

7.2.21
Ligaments of ankle joint, lateral view.

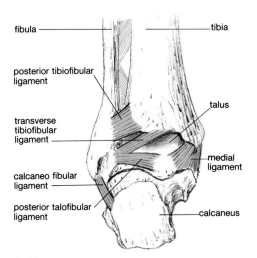

7.2.22
Ligaments of ankle joint, posterior view.

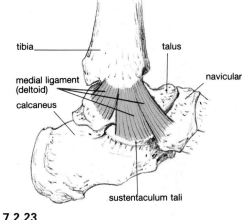

7.2.23
Ligaments of ankle joint, medial view.

THE TIBIO-FIBULAR JOINTS

The head of the fibula articulates with the postero-lateral aspect of the lateral condyle of the tibia. This **superior tibio-fibular** joint (**7.2.16**) is a plane synovial joint which permits the fibula to rotate slightly as the talus moves in the ankle joint (see below). The shaft of the tibia is united for most of its length to the fibula by an interosseous membrane which serves principally for the attachment of muscles. The lower ends of the shafts of the tibia and fibula are united by a fibrous **inferior tibio-fibular** joint at which little movement can occur (see **7.2.20**). Anterior and posterior ligaments and a deep transverse tibio–fibular ligament reinforce this joint and contribute to the socket into which the wedge-shaped superior articular surface of the talus fits.

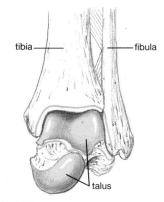

7.2.20
Anterior attachments of capsule of left ankle joint, and inferior tibiofibular joint.

THE ANKLE JOINT

A. Living anatomy

Examine the range of movement of your ankle. When standing the foot is at right angles to the leg. It can be flexed (plantar-flexed) by about 60° and extended (dorsiflexed) by only about 15°. Examine the skeleton and note that the superior articular surface of the talus, which articulates with the tibia and fibula, is wider anteriorly than posteriorly. Therefore, when the foot is extended (as in walking up hill) the ankle is more stable than when flexed (as in walking down hill), when a small amount of abduction and adduction is possible. Now move your foot and discover whether other movements occur at the ankle. Inversion and eversion of the foot occur only at the subtalar and transverse tarsal joints (see Seminar 6).

B. Prosection

The capsule of the ankle joint (**7.2.20**) is attached around the articular margins of the tibia, fibula, and talus. As in other hinge joints, the capsule is reinforced by strong lateral and medial ligaments. The **lateral ligament** (**7.2.21**) has three parts, each of which may be separately damaged in a 'sprain' of the joint. The **anterior talo-fibular** part is attached superiorly to the lateral malleolus and runs forward and medially to the neck of the talus; the **calcaneo-fibular** part runs downward and backward from the tip of the malleolus to the lateral side of the calcaneum; the **posterior talo-fibular** part (**7.2.22**) runs horizontally from the malleolar fossa on the inner aspect of the medial malleolus to the posterior aspect of the talus. The fan-shaped **medial (deltoid) ligament** (**7.2.23**) is attached above to the medial malleolus of the

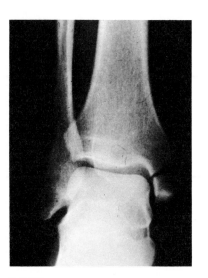

7.2.24
Fractured medial malleolus (compare with
7.2.25).

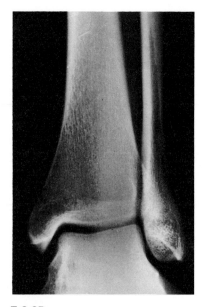

7.2.25
Ankle joint (AP).

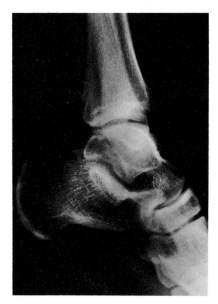

7.2.26
Ankle, subtalar, and mid-tarsal joints
(lateral).

tibia and below to the navicular and the sustentaculum tali of the calcaneus. It is very strong compared to the lateral ligament. Such is the strength of the medial ligament that, when the talus is twisted in its mortice to the extent that ligaments or bones give way, it is more common for the medial malleolus to be pulled off the shaft of the tibia than for the medial ligament to rupture. Such a fracture is shown in **7.2.24**.

Examine A–P and lateral radiographs of the ankle joint (**7.2.25, 7.2.26**) and note that:

● in the adult there is overlap between the tibia and fibula at the inferior tibio-fibular joint. In a normal joint it is not possible to see between them; if you can, then a fracture of the joint has occurred.

● the ankle joint space is uniform. The maximum distance between the medial surface of the talus and the lateral articular surface of the medial malleolus should be about the same as that between the lateral surface of the talus and the articular surface of the fibula. If it is much greater, then the talus has moved laterally and the medial ligament must have ruptured.

● in the radiograph of a young patient (see **7.1.16**), the epiphysial plate of the lower end of the fibula lies at the same level as the line of the ankle joint, i.e. at a lower level than that of the tibia. Note the plaster casts on both feet.

For an MRI of the ankle joint see **7.6.4**.

Qu. 2F   *If a patient complained of pain in the left ankle after 'twisting' it by falling off a kerb, what structures might have been damaged?*

After you have completed the next three seminars make a list of the groups of muscles which act on the hip, knee, and ankle joints.

The subtalar joint and other joints in the foot are considered in Seminar 6.

Requirements:

Articulated skeleton
Prosections of hip, knee, and ankle joints
Radiographs of hip, knee, and ankle joints.

Seminar 3

Gluteal region and muscles moving the hip joint

Aims

To study the ligaments and muscles of the pelvic girdle which are concerned with maintenance of the upright stance, the transmission of trunk weight to the lower limbs, and the mechanisms of gait (Chapter 9); also the emergence of blood vessels and nerves from the pelvis which supply the skin and muscles of the gluteal region, the back of thigh and leg.

A. Living anatomy

The iliac crest demarcates the gluteal region from the abdomen. Run your fingers backward from the anterior superior iliac spine to the posterior superior iliac spine which lies under a dimple on the skin of the back opposite the spinous process of S2. The mass of the buttock is composed of muscle (gluteus maximus) and is limited below by the crescentic-shaped fold of the buttock. Sit down on your hands and feel the bones on which you sit, the ischial tuberosities. Keeping your hands in place, stand up and sit down and note how, when you sit down, gluteus maximus appears to rise and uncover the ischial tuberosities.

With your hands over gluteus maximus, extend your lower limb against resistance; at what degree from the vertical does this muscle contract? Step up on to a small stool and again note the contraction of gluteus maximus extending the hip joint.

Stand with both feet together on the ground and palpate the muscles just below the anterior part of both iliac crests. These are tensor fasciae latae and gluteus medius; gluteus minimus, which has a similar action, lies more deeply. Now lift one leg off the ground. As you do so, the pelvis is tilted toward the supporting leg by the contraction of these muscles, which are abducting the pelvis on the femur; palpate these muscles on both sides and walk slowly forward. Note the contraction of muscles and the slight tilt of the pelvis that occurs with each stride; these move the centre of gravity over the supporting leg and help the non-weight-bearing limb to swing free of the ground.

Remind yourself of the range of flexion of the hip joint; the major muscles responsible for this and for lateral rotation lie deeply and cannot be palpated. The muscles responsible for adduction and medial rotation will be studied in Seminar 4.

B. Prosections

Gluteal region

If you are dissecting rather than using prosections, the skin incisions you should make are shown in **7.3.1**.

Before studying the muscles of the gluteal region examine the bony pelvis with its attached ligaments (**7.3.2**). Locate the strong **sacrotuberous ligament** which attaches the side of the sacrum to the ischial tuberosity, and the **sacrospinous ligament** which lies deep to the sacrotuberous ligament and is attached to the side of the sacrum and the ischial spine. Because of the slope of the sacrum within the pelvis, and the transmission of weight from the spine on to its upper surface, the sacrum tends to rotate anteriorly within the pelvis (see **9.1**). The strong sacrotuberous ligament, coupled with the strength of the sacroiliac joint (p. 152), prevents this rotation.

Gluteus maximus (**7.3.3**) is a coarsely fasciculated muscle that arises in a continuous line from the posterior part of the ilium, the lumbar fascia,

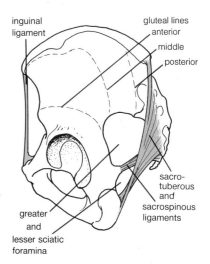

7.3.2
Sacrotuberous and sacrospinous ligaments.

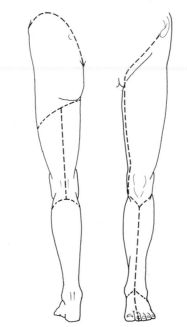

7.3.1
Incision lines for dissection of lower limb.

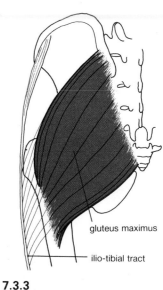

7.3.3
Gluteus maximus.

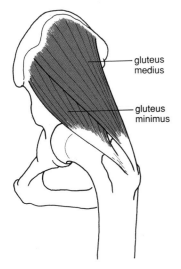

7.3.4
Gluteus medius and minimus.

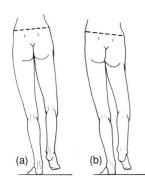

7.3.5
(a) Normal pelvic tilt and
(b) Trendelenberg sign.

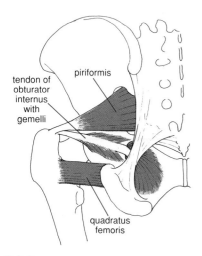

7.3.6
Piriformis, internal oblique and
gemelli, quadratus femoris.

and the sides of the sacrum and coccyx; it also arises from the sacrotuberous ligament which attaches between the side of the sacrum and the sacral aspect of the ischial tuberosity. Gluteus maximus is quadrilateral in shape. It passes obliquely downward and laterally to insert into the gluteal tuberosity, a roughened area on the back of the femur below the lesser trochanter. It also inserts into the **ilio-tibial tract**, a thickening of the investing deep fascia of the leg (fascia lata) which forms a band running from the level of the greater trochanter to the lateral condyle of the tibia and into which tensor fasciae latae (a muscle arising from the iliac crest) is inserted. Examine the innervation and blood supply of gluteus maximus from the inferior gluteal nerve and inferior gluteal vessels. The neurovascular bundle emerges from the pelvis inferior to piriformis (see below). Gluteus maximus also receives a blood supply (but not innervation) from the superior gluteal neurovascular bundle.

Qu. 3A *When studying the living body you will already have noted that gluteus maximus extends the trunk on the thigh, or conversely the thigh on the trunk, and is a prime mover in getting you out of your chair! What other action has this muscle?*

Reflect gluteus maximus to uncover **gluteus medius (7.3.4)** which arises from a large area of the lateral surface of the ilium (between the posterior and middle gluteal lines) and inserts into the lateral aspect of the greater trochanter of the femur; and **gluteus minimus (7.3.4)** which arises from the iliac blade deep to gluteus medius (between the middle and anterior gluteal lines) and inserts into the anterior aspect of the greater trochanter. Locate **tensor fasciae latae** (see **7.4.3**) at its origin from the anterior part of the iliac crest and trace it to its attachment into the thick ilio-tibial tract of the fascia lata. Through this it inserts into the lateral condyles of the femur and tibia. All three of these muscles are supplied by the **superior gluteal nerve** in the neurovascular bundle which emerges from the pelvis above piriformis.

Remind yourself of the action of these muscles to support and tilt the pelvis during walking. If the gluteal muscles are paralysed or the hip is (congenitally) dislocated, the abduction mechanism will no longer function. When a patient stands on a leg which is affected in one of these ways, instead of the opposite side of the pelvis rising, it will fall. The patient is said to have a positive Trendelenburg sign **(7.3.5)**. Also, the patient's gait is characteristically waddling and is described as a Trendelenburg gait.

The sacrotuberous and sacrospinous ligaments create two foramina into the pelvis, one above the sacrospinous ligament—the **greater sciatic foramen**; the other below it—the **lesser sciatic foramen**. Identify the structures which pass through the greater sciatic foramen. The most prominent is **piriformis (7.3.6)** which arises from the anterior aspect of the sacrum, passes through the greater sciatic foramen into the gluteal region, runs laterally, posterior to the hip joint, and inserts

into the upper border of the greater trochanter under cover of gluteus medius. The **superior gluteal vessels** branch from the internal iliac vessels within the pelvis, and the **superior gluteal nerve** originates from L4, 5 and S1 in the lumbosacral plexus. Both these structures emerge from the pelvis above piriformis. The large **sciatic nerve** (L4, 5, S1, 2, 3) which also originates from the sacral plexus, passes through the greater sciatic foramen below piriformis and enters the lower, inner quadrant of the buttock, separated from the posterior aspect of the hip joint by only the small lateral rotator muscles **(7.3.7)**. It is therefore at risk in posterior dislocations of the hip joint. The **posterior cutaneous nerve of the thigh** (which is distributed as its name implies) lies superficial to the sciatic nerve, whereas the **inferior gluteal nerve** and blood vessels, which also originate from the lumbosacral plexus (L5, S1,2) and internal iliac vessels respectively, pass directly into gluteus maximus **(7.3.7)**.

Examine a bony pelvis in which the sacrotuberous and sacrospinous ligaments are attached and then locate, on a prosected part, the ischial spine and its attached sacrospinous ligament. Having done so, examine the **pudendal nerve**, the **internal pudendal artery**, and the **nerve to obturator internus**, all of which arise in the pelvis from either the sacral plexus or internal iliac artery. They leave the pelvis by the greater sciatic foramen, cross the gluteal aspect of the sacrospinous ligament, and then enter the pelvis or perineum via the lesser sciatic foramen. The pudendal nerve and vessels supply the perineal region. The nerve to **obturator internus** supplies the muscle as it takes origin from the walls and fascia of the obturator foramen within the pelvis. Its tendon emerges from the lesser sciatic foramen, runs laterally deep to the sciatic nerve, thereby separating it from the hip joint, and inserts into the greater trochanter. Although it appears to be fleshy this is because it is flanked by the **superior and inferior gemelli** which arise from the bony margins of the lesser sciatic foramen and insert into the tendon of the obturator internus.

Identify **quadratus femoris** which runs horizontally from the lateral border of the ischial tuberosity to the intertrochanteric crest of the femur. Its nerve emerges from the greater sciatic foramen deep to the sciatic nerve and enters the deep aspect of the muscle.

Qu. 3B *What is the action of piriformis, obturator internus, and quadratus femoris?*

Now turn to a prosection of the muscles on the anterior aspect of the hip joint **(7.3.8)**. Identify **psoas major** which arises from the bodies and transverse processes of lumbar vertebrae, and **iliacus** which arises from the iliac fossa. Both these muscles emerge from the pelvis deep to the inguinal ligament (which runs from the anterior superior iliac spine to the pubic tubercle), pass over the anterior aspect of the hip joint (the tendon of psoas is separated from the joint by a bursa), and insert into the lesser trochanter of the femur. Some fibres of iliacus insert into the psoas tendon. Psoas major

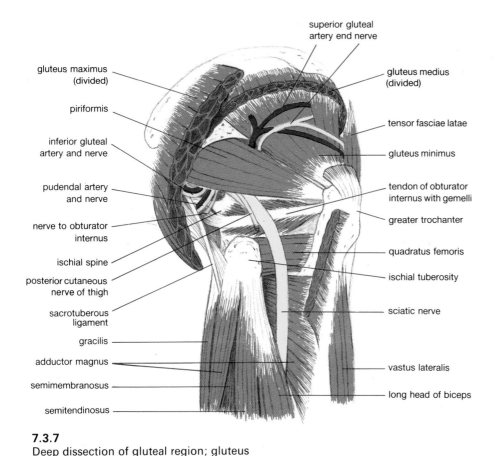

gluteus maximus (divided)

piriformis

inferior gluteal artery and nerve

pudendal artery and nerve

nerve to obturator internus

ischial spine

posterior cutaneous nerve of thigh

sacrotuberous ligament

gracilis

adductor magnus

semimembranosus

semitendinosus

superior gluteal artery end nerve

gluteus medius (divided)

tensor fasciae latae

gluteus minimus

tendon of obturator internus with gemelli

greater trochanter

quadratus femoris

ischial tuberosity

sciatic nerve

vastus lateralis

long head of biceps

7.3.7
Deep dissection of gluteal region; gluteus maximus and medius largely removed.

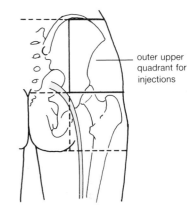

psoas major

iliacus

pectineus

7.3.8
Psoas, iliacus, and pectineus.

outer upper quadrant for injections

7.3.9
Site for injections in the buttock.

is supplied segmentally by spinal nerves; iliacus by the femoral nerve.

Locate the small muscle **pectineus** which arises from the pectineal surface of the pubis and passes distally to insert just below the lesser tuberosity. It is a weak flexor of the hip joint and is supplied by the femoral nerve.

Qu. 3C *What are the actions of psoas and iliacus on the hip joint?*

Qu. 3D *In what way would a fracture of the neck of the femur alter the action of psoas and iliacus?*

The muscles producing adduction and medial rotation of the hip are situated in the thigh and will be studied in the next seminar.

The buttock provides a large muscle mass which may be used for deep injections; they should be given into the upper outer quadrant of the buttock to avoid damage to the sciatic nerve (**7.3.9**). The deltoid or quadriceps is more commonly used for intramuscular injections.

Finally, consider the danger of a posterior disloca-

tion of the hip joint. If a passenger, not wearing a seat belt, is sitting in a car which is brought to a sudden halt, the passenger is projected forwards. With the hip flexed to a right angle and the full force of the blow taken on the knees, the head of the femur may be pushed backward out of the acetabulum, quite commonly taking with it a sharp fragment of the lip of the acetabulum. Since the sciatic nerve runs across the posterior aspect of the acetabulum, it may very easily be bruised or even lacerated. As you continue to study the lower limb, consider the consequences of such damage.

Requirements:

Articulated skeleton
Pelvis with sacrotuberous and sacrospinous ligaments attached
Prosections of gluteal region showing gluteus maximus in place, and gluteus maximus reflected to reveal the deep muscles of the gluteal region, the superior and inferior gluteal neurovascular bundles, and the sciatic and pudendal nerves
Prosections of the hip joint to show psoas major, iliacus, and pectineus.

Seminar 4

Muscles of the thigh and movements of the knee

Aims To study the various groups of muscles in the thigh which are responsible for the support and movements of the hip and knee joints.

A. Living anatomy

Palpate the antero-medial aspect of your thigh just below the groin to locate the tendon of **adductor longus**. Follow the tendon up to its origin on the body of the pubic bone immediately below and medial to the pubic tubercle. Now forcibly adduct your hip joint and feel the muscle contract. Stand up and, while feeling the tendon, medially and laterally rotate your extended lower limb against resistance.

Qu. 4A *Does the muscle contract in medial or lateral rotation?*

Extend your knee joint against resistance while palpating the anterior aspect of your thigh to feel the contraction of **quadriceps femoris**, the homologue of triceps in the arm. Note the muscular bulges on either side of the patella; that on the medial aspect is the more distal and more pronounced. Now sit down, palpate the mid-point of the groin, and lift your leg from the floor. As you flex your hip you will feel the contraction of **rectus femoris**, a part of quadriceps.

Wasting of quadriceps is an early physical sign of disease of the knee joint. This is often apparent first in the prominence of vastus medialis above the medial aspect of the knee. On your partner compare the size of the muscular prominence on both sides and see if you detect any difference. Also measure the circumference of the thighs at two different levels above the patella. Often the dominant thigh (as for kicking) will be measurably greater than the other.

Elicit a **knee jerk**: get your partner to sit down and relax, with one leg crossed over the other. Tap the quadriceps tendon of the crossed leg with a patellar hammer. Now compare the extent and 'quality' of the knee jerk with that of the opposite side. Record your results and explain the mechanisms underlying the reflex response.

Stand up and extend your knee fully. As full extension is approached try to feel the slight medial rotation of the femur on the tibia that occurs as the knee joint 'locks'. Before flexion can occur this rotation has to be reversed by the contraction of popliteus.

Palpate the 'hamstrings' on the posterior aspect of your thigh just above the knee joint. You should be able to feel two tendons (**semimembranosus** and **semitendinosus**) medially, and a single tendon (**biceps femoris**) laterally (7.4.8, 7.4.9 and see 7.5.3). Now flex your knee joint and then extend your hip joint against resistance.

Qu. 4B *In which of these movements do the hamstrings contract?*

B. Prosections

Before you study the muscles of the thigh, note the thick layer of deep fascia (**fascia lata**) that surrounds the entire thigh like a sleeve. It is pierced by some veins and nerves and is thickened laterally to form the ilio-tibial tract (p. 108, see **7.4.3**).

Muscles of the front of thigh (7.4.1)

First, review psoas major, iliacus, and pectineus (p. 109), which act on the hip joint.

Identify the four parts of the **quadriceps femoris** (7.4.2): **rectus femoris** originates from the anterior inferior iliac spine and the ischium immediately above the acetabulum; **vastus intermedius** arises from the front and sides of the femoral shaft; **vastus medialis** arises in a continuous line from the medial aspect of the intertrochanteric line and the linea aspera; and **vastus lateralis** arises from the lateral aspect of the intertrochanteric line, the base of the greater trochanter, and the linea aspera. All four muscle bellies insert into the **quadriceps tendon** which inserts into the upper margin of the patella and, via the patella and **patellar tendon**, to the tibial tuberosity. Vastus medialis has an additional attachment to the medial surface of the patella.

These knee extensors are all supplied by the femoral nerve (L2,3,4).

Qu. 4C *Which muscle of the quadriceps group acts on the hip joint and what movement is produced?*

Qu. 4D *Why is the attachment of vastus medialis to the medial surface of the patella so important?*

Examine **sartorius** (7.4.3) which arises from the anterior superior iliac spine and crosses the thigh obliquely to insert into the anterior aspect of the tibia below the medial condyle. Sartorius (from the

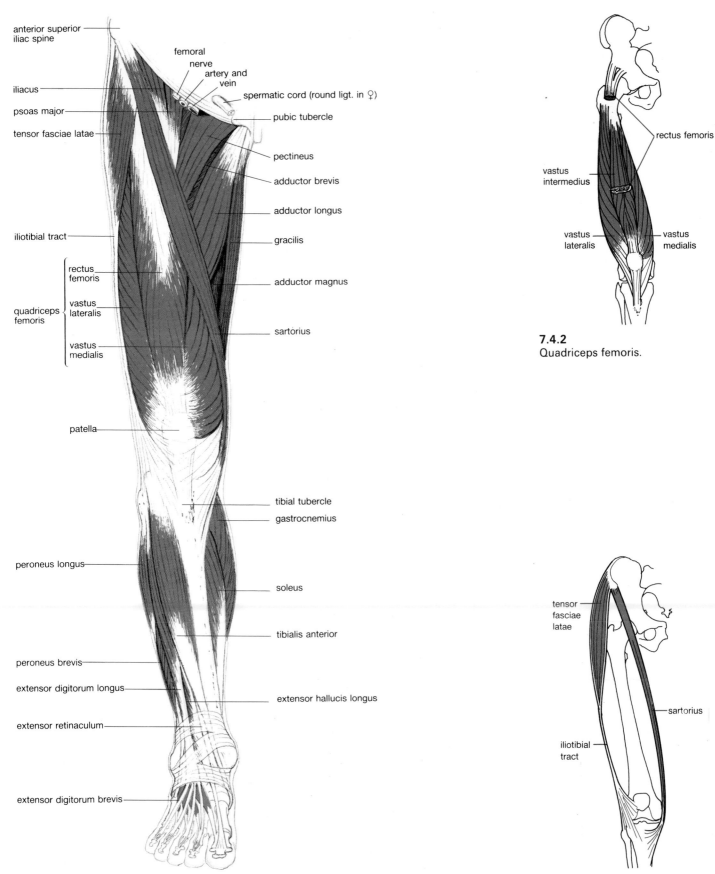

anterior superior iliac spine

femoral
nerve
artery and vein

iliacus

spermatic cord (round ligt. in ♀)

psoas major

pubic tubercle

tensor fasciae latae

pectineus

adductor brevis

adductor longus

iliotibial tract

gracilis

rectus femoris

adductor magnus

quadriceps femoris

vastus lateralis

sartorius

vastus medialis

patella

tibial tubercle

gastrocnemius

peroneus longus

soleus

tibialis anterior

peroneus brevis

extensor digitorum longus

extensor hallucis longus

extensor retinaculum

extensor digitorum brevis

7.4.1
Muscles of the front of the lower limb.

rectus femoris

vastus intermedius

vastus lateralis

vastus medialis

7.4.2
Quadriceps femoris.

tensor fasciae latae

sartorius

iliotibial tract

7.4.3
Sartorius, tensor fasciae latae.

Latin word for tailor) moves the thigh into the cross-legged sitting position. Like quadriceps and iliacus (p. 109), it is supplied by the femoral nerve.

Look again at tensor fasciae latae (**7.4.3**) as it arises from the anterior part of the iliac crest and passes down to insert into the ilio-tibial tract; remind yourself of its functions (pp. 107, 108).

The extensor mechanism of the knee

The quadriceps femoris, the patella, and the patellar ligament are together referred to as '**the extensor mechanism of the knee**', to emphasize the unity and function of the separate structures. The stability of the knee depends on it. If quadriceps is weakened, the knee will very easily give way, causing the body to fall. If the extensor mechanism is subjected to excessive strain the tibial tuberosity can be avulsed from its bed or the lower pole of the patella pulled off, and the patella may fracture or the retinacula tear. Such injuries usually occur in young age groups.

In the elderly, however, the quadriceps tendon is more likely to give way, so that the patient cannot extend the knee fully against gravity.

> **Qu. 4E** *Examine 7.4.4. The patient complained of pain in the thigh during a rugby match. What has happened?*

The line of pull of quadriceps muscle is essentially in line with the shaft of the femur. Since the femur is not vertical but the tibia is, quadriceps tends to pull the patella a little laterally. This tendency is counteracted by the lower fibres of the vastus medialis which insert into the upper medial edge of the patella. Also, the lateral condyle of the femur projects more anteriorly than the medial condyle, reducing the tendency to lateral displacement of the patella. Sometimes, however, if the patella is smaller and slightly higher than normal, or if there is a tendency to 'knock knee', the patella can suffer recurrent dislocation, moving laterally to the extent that it jumps over the lateral condyle. When this occurs, quadriceps is reflexly inhibited and the patient's knee 'collapses'.

Muscles of the medial aspect of thigh (7.4.5, 7.4.6, 7.4.7)

The muscles on the medial aspect of the thigh all adduct the hip. **Pectineus** (p. 109) arises from the superior pubic ramus and inserts into the femur below the lesser trochanter; **adductor longus** and **adductor brevis** also arise from the pubic bone (adductor longus by a tendon attached immediately below the pubic tubercle; brevis from the body and inferior ramus of the pubis) and pass medially and downward to insert into the linea aspera. **Gracilis** runs from the body of the pubis to the medial aspect of the tibial shaft at its upper end, just posterior to sartorius.

Adductor magnus has an adductor part which arises from the ischio-pubic ramus and a hamstring part which arises from the ischial tuberosity. Its fibres pass medially and downward to insert into the whole length of the linea aspera and the upper part of the medial supracondylar line. At this point it forms a fibrous arch through which the femoral neurovascular bundle passes from the extensor (anterior) to the flexor (posterior) compartment of the thigh. The distal tendinous end of the arch is attached to the adductor tubercle.

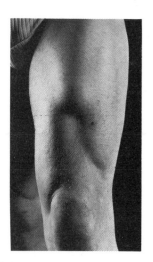

7.4.4
See Qu. 4E.

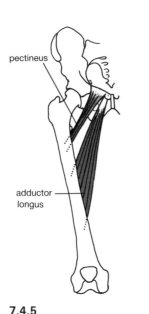

7.4.5
Pectineus, adductor longus.

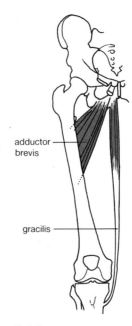

7.4.6
Adductor brevis, gracilis.

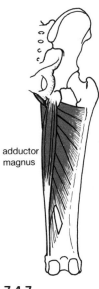

7.4.7
Adductor magnus.

Qu. 4F *What actions other than adduction might any of these muscles perform?*

The entire adductor muscle group is supplied by the **obturator nerve** (L2,3,4) which leaves the pelvis via the obturator foramen, and supplies and pierces **obturator externus** which is attached to its outer bony margins. The tendon of the obturator externus winds laterally below and behind the neck of the femur to insert into the medial aspect of the greater trochanter; it can therefore rotate the femur laterally.

Qu. 4G *If you needed to locate the pubic tubercle accurately (to determine whether a hernia was of the inguinal or femoral variety), how, with a knowledge of the muscles you have just studied, could you do so?*

Muscles of the back of the thigh
(7.4.8, 7.4.9)

The group of muscles which lie in the posterior compartment of the thigh are known as the 'hamstrings'. They all flex the knee joint and most also extend the hip. They all arise from the ischial tuberosity. **Semimembranosus** originates by a 'membranous' aponeurosis and inserts into the grooved posterior aspect of the medial condyle of the tibia; **semitendinosus**, as its name suggests, ends in a long tendon which passes downward superficial to semimembranosus and then on the medial aspect of the knee to insert into the upper end of the medial aspect of the tibial shaft behind sartorius and gracilis. The **long head of biceps femoris** joins its **short head** which arises from the linea aspera and the lateral supracondylar ridge of the femur; the combined muscle ends in a tendon which passes downward to insert

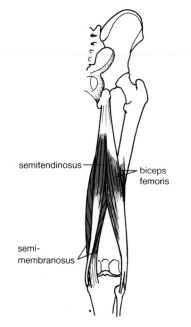

7.4.9
Hamstring muscles.

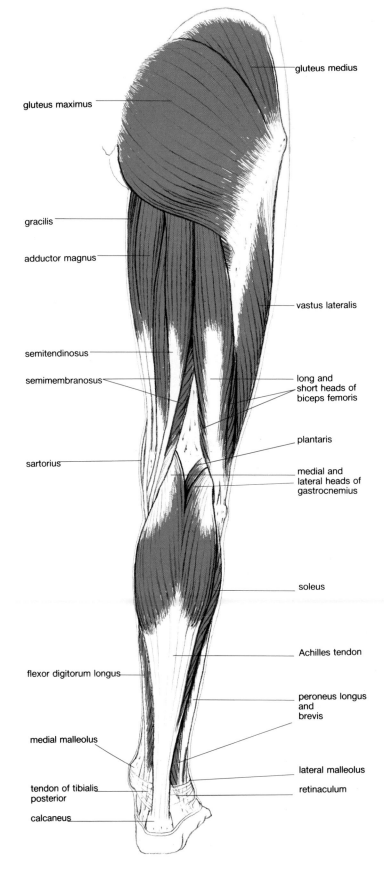

7.4.8
Muscles of the back of the lower limb.

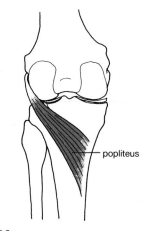

7.4.10
Popliteus.

into the head of the fibula on either side of the lateral ligament of the knee joint. That part of adductor magnus which arises from the ischial tuberosity is also considered as a hamstring. The hamstrings are all supplied by the **sciatic nerve** (L4,5, S1,2,3).

Qu. 4H *Which nerves supply adductor magnus?*

Examine a prosection which shows **popliteus** (**7.4.10, 7.2.18**), a small triangular muscle situated deeply in the upper part of the calf. It arises on the posterior surface of the tibia above the soleal line and its tendon passes into the knee joint below the lower extremity of the capsule. Within the knee joint its tendon inserts partly into the posterior aspect of the lateral meniscus and partly into a pit on the outer aspect of the lateral condyle of the femur. Note that the tendon separates the lateral meniscus from the lateral ligament of the knee, and that it lies in a groove on the lateral condyle when the knee is flexed. It is supplied by the tibial nerve.

Qu. 4I *What is the function of popliteus?*

'Guy-ropes' of the pelvis

Consider the attachments of sartorius, gracilis, and semitendinosus. All insert into the medial aspect of the upper end of the tibia, but their origins on the pelvis are separated very widely, from the anterior superior iliac spine, the pubis, and the ischial tuberosity respectively. These long parallel-fibres muscles are thought to act like guy-ropes on a tent, stabilizing the pelvis in relation to the tibia.

Requirements:

Articulated skeleton and separate lower limb bones
Prosections of anterior, medial, and posterior compartments of the thigh
Patellar hammer.

Seminar 5

Muscles of the leg and movements of the ankle and foot

Aims To study the muscles of the different compartments of the leg and the movements of the ankle and foot that they produce.

A. Living anatomy

Palpate the front of your leg to feel the sharp anterior margin of the tibia (shin); trace it upward and downward and note that the subcutaneous surface of the tibia extends from the knee to the medial malleolus. Now palpate the lateral malleolus and try to trace the fibula upward; it is largely encased in muscle. Palpate the muscle group between the tibia and fibula anteriorly and note its contraction when the ankle is extended (dorsiflexed), when the foot is inverted, and when the toes are extended against resistance (**7.5.1**). At the front of the ankle identify the tendons of **tibialis anterior** medially and **extensor hallucis longus** and **extensor digitorum longus** more laterally. Palpate the area immediately lateral to the tendon of extensor digitorum longus and again extend your toes against resistance to feel the contraction of a small muscle—**extensor digitorum brevis**.

Evert your foot against resistance (**7.5.2**). Feel for the contraction of the **peroneal** muscles which overlie the fibula and note their tendons running downward behind the lateral malleolus.

Get your partner to stand and take his weight on one leg, rising on to the toes. In the calf the two bellies of **gastrocnemius** stand out superficially, and **soleus** bulges beneath and on either side of them, especially distally (**7.5.3**). Trace both muscles downward into the **Achilles tendon** and note its attachment to the calcaneus at the heel.

To elicit an **ankle jerk**, hold your partner's foot in slight dorsiflexion, tap the Achilles tendon with a tendon hammer, and note the reflex contraction of gastrocnemius and soleus. This reflex tests the integrity of the S1 segment of the spinal cord.

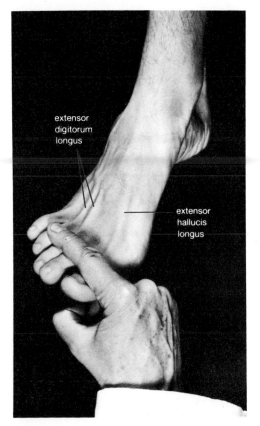

extensor digitorum longus

extensor hallucis longus

7.5.1
Extension of toes against resistance.

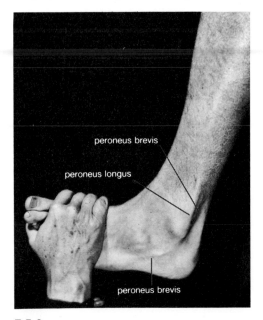

peroneus brevis

peroneus longus

peroneus brevis

7.5.2
Eversion of foot against resistance.

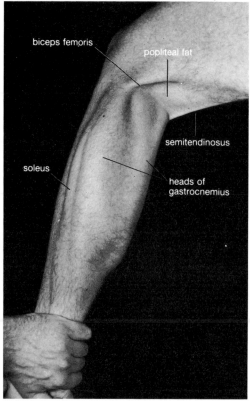

7.5.3
Flexion of knee against resistance. This outlines the diamond-shaped popliteal fossa which lies posterior to the knee between the hamstring tendons and the two muscular heads of gastrocnemius.

B. Prosections

Muscles of the front of leg and dorsum of foot

Tibialis anterior, extensor digitorum, and extensor hallucis longus (see **7.4.1**) all arise from the anterior aspect of the shaft of either the tibia or the fibula and the interosseous membrane. The most medial is **tibialis anterior (7.5.4)**. Trace its tendon to its attachment on the medial aspect of the medial cuneiform bone and the adjoining part of the base of the first metatarsal. Identify the tendon of **extensor hallucis longus (7.5.4)** at its insertion into the base of the distal phalanx of the big toe and trace it back to its muscle belly. The tendons of **extensor digitorum longus (7.5.5)** lie more laterally on the dorsum of the foot; they broaden into extensor expansions on the lateral four toes. As in the hand, the central part of each tendon attaches to the base of the 2nd phalanx while the lateral parts insert into the base of the distal phalanx. An extension of this muscle, known as peroneus tertius, inserts into the dorsal aspect of the 5th metatarsal. **Extensor digitorum brevis (7.5.5)** arises from the dorsal surface of the calcaneum and fills what otherwise would be a hollow on the lateral aspect of the dorsum of the foot. Three of its tendons join the dorsal extensor expansion while that to the big toe (sometimes termed extensor hallucis brevis) has a separate insertion into the base of the proximal phalanx. All this muscle group is supplied by the **deep peroneal nerve**, the 'extensor' component of the sciatic nerve.

The anterior compartment is closed by a very firm fascia, and swelling of its contents can cause severe pressure effects (see p. 127). At its distal end

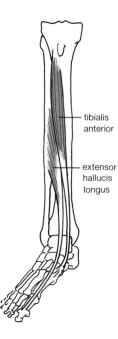

7.5.4
Tibialis anterior, extensor hallucis longus.

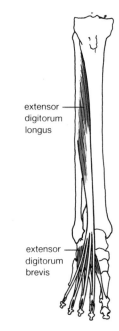

7.5.5
Extensor digitorum longus, extensor digitorum brevis.

the fascia forms an **extensor retinaculum** which keeps the tendons in place on the front of the ankle (see **7.6.11**).

Qu. 5A *What are the actions of these muscles, in particular those of tibialis anterior?*

Muscles of the lateral aspect of leg

The lateral compartment of the leg contains **peroneus longus (7.5.6)** which arises from the upper two-thirds of the lateral surface of the fibula and **peroneus brevis** which lies beneath peroneus longus and arises from the lower two-thirds of the same bony surface. The tendon of peroneus longus runs behind the lateral malleolus, and then grooves the cuboid bone and passes beneath the sole of the foot to insert on the medial cuneiform bone and the base of the 1st metatarsal close to tibialis anterior **(7.6.5)**. The tendon of peroneus brevis also passes behind the lateral malleolus; it inserts into the base of the 5th metatarsal. The **peroneal retinaculum** holds the tendons in place behind the lateral malleolus. Peroneus longus and brevis are supplied by the **superficial peroneal nerve**.

Qu. 5B *What actions do these muscles have?*

Muscles of the calf

The superficial calf muscles (see **7.4.8**) are gastrocnemius and soleus **(7.5.7)**. **Gastrocnemius** arises by two fleshy bellies, one from each of the condyles of the femur, which unite to insert by the **tendo calcaneus** or **Achilles tendon** into the posterior surface of the calcaneus. **Soleus** arises from a horseshoe-shaped origin which spans the tibia and fibula below popliteus, its tendon blending with the tendo calcaneus. Perforating veins pass through soleus and it is therefore an important component of the venous pump (p. 129).

Qu. 5C *What are the actions of the superficial calf muscles?*

The Achilles tendon transmits very considerable forces, and is the thickest tendon in the body. In middle age, if a person suddenly takes part in relatively unaccustomed exercise, the tendon may snap, usually at the site of some degenerative changes. The typical story is of a middle-aged man or woman starting to play badminton again after a long lay-off. He suddenly feels as if his partner has hit him over the heel, and he may even hear the snap, 'like a pistol shot!'

Qu. 5D *What effect would such an injury have on the function of the foot*
a) when free of the ground?
b) when weight-bearing?

The integrity of the Achilles tendon can be tested by squeezing the calf from side to side with the ankle relaxed ('squeeze test'). If the Achilles tendon is intact the foot moves passively into plantar flexion at each squeeze and returns to the resting position when the pressure is relaxed. If the tendon is ruptured no such movement occurs. A ruptured Achilles tendon should be resutured as soon as possible to avoid its healing in a lengthened state, which would weaken the 'push–off' of the foot in walking.

Deep muscles of the calf (7.5.8)

Lying deeply under the superficial muscles are the **flexor hallucis longus (7.5.9)**, **flexor digitorum longus (7.5.9)**, and **tibialis posterior (7.5.10)**, which take origin, respectively, from the posterior aspect of the fibula, the tibia, and the interosseous membrane and surrounding bone.

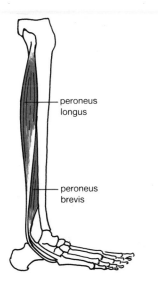

7.5.6
Peroneus longus, peroneus brevis.

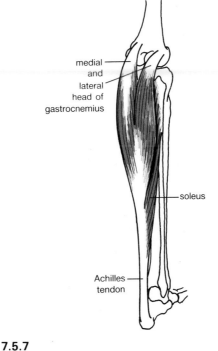

7.5.7
Gastrocnemius, soleus.

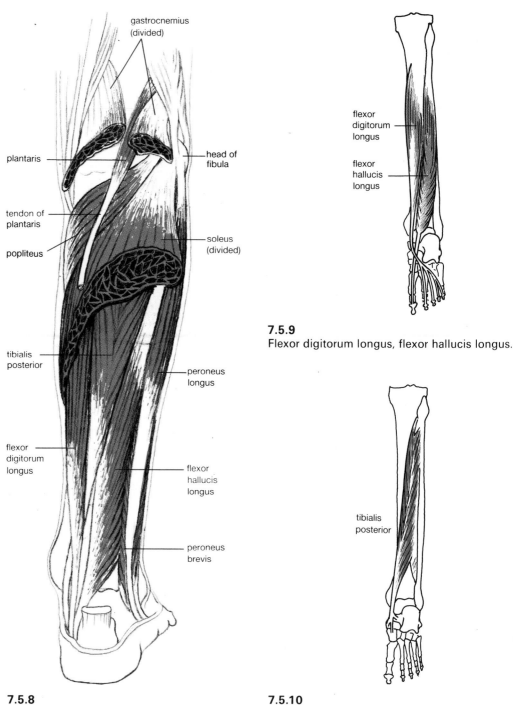

7.5.8
Deep muscles of the calf.

7.5.9
Flexor digitorum longus, flexor hallucis longus.

7.5.10
Tibialis posterior.

They insert, respectively, into the distal phalanx of the hallux, into the distal phalanges of the lateral four toes, and into the navicular and adjacent tarsal bones.

At the ankle joint they are held in position by the **flexor retinaculum** (see **7.6.6**) which is attached to the medial malleolus, to the sustentaculum tali, and to the medial surface of the calcaneus below the sustentaculum.

All muscles of the flexor compartment of the calf are supplied by the **tibial nerve**, the 'flexor' component of the sciatic nerve.

For an MRI showing the ankle joint and Achilles tendon see **7.6.4.**

Requirements:

Articulated skeleton and separate bones of the lower limb
Prosections of the anterior compartment of the leg and dorsum of the foot; of the lateral compartment of the leg; of the superficial muscles of the calf; and of the deep muscles of the calf.

Seminar 6

Joints, muscles, and movements of the foot

Aims To study the feet. To consider their role in the transmission of body weight to the ground, and the bony, muscular, and ligamentous components of the arches which enable them to provide a flexible and resilient base on which standing, and all forms of gait, depend; to compare functional adaptations of the foot with those of the hand.

A. Living anatomy

The feet are subject to enormous stresses and strains and are therefore frequently injured. Such injuries can cause prolonged pain and difficulties with walking.

Examine first the position of the feet. Do they face forward or are they angled laterally or medially ('pigeon-toed')?

Qu. 6A *If either or both feet is/are laterally or medially rotated, to which joint(s) of the lower limb can this position be attributed?*

Ask your partner to remove his shoes and socks, and to stand relaxed with his feet slightly apart.

Examine the shape of the feet. The medial aspect of the foot should form a distinct, if flattened, longitudinal **arch** between the heel and the meta-tarsal heads (**7.6.1**); examine other feet and note any differences in the height of this arch. Examine the lateral aspect of your partner's foot. The skin is probably in contact with the ground, but this conceals a shallow lateral arch in the bones. Finally examine the sole and note the transverse curvature which forms half of an arch. Look at the heel from behind, noting that it is somewhat everted. Now watch the medial aspect of the longitudinal arch and the heel as your partner stands on 'tip-toe'. The heel moves from slight eversion to slight inversion.

Qu. 6B *Does the longitudinal arch increase or decrease?*

Note any angulation of the great toe on the foot. Its metatarso-phalangeal joint is commonly angled laterally and enlarged to form a 'bunion' (**7.6.2**); some of these deformed joints require operative reshaping.

Examine the skin of the foot, noting the natural thickenings over the heel and pads of the toes, and any other thickened areas ('corns') resulting from ill-fitting shoes. Examine the toes for any inflammation caused by 'ingrowing' toenails; another consequence of wearing shoes. Examine the skin webs

7.6.1
Longitudinal arch of the foot.

between the toes; they can vary considerably in extent (p. 28).

Ask your partner to flex and extend, and then invert and evert his foot.

Qu. 6C *At which joints do the movements occur?*

Qu. 6D *What movements can be made at the metatarso-phalangeal joints and the interphalangeal joints?*

Ask your partner to stand up and take a step forward. Notice how, as the heel of the weight-bearing foot rises from the ground, the big toe is passively dorsiflexed at the metatarso-phalangeal joint. This stretches flexor hallucis longus, before its contraction, which provides the final thrust-off in the striding gait (p. 154); it also stretches the plantar ligaments of the foot, thereby tensing the arch.

B. Prosections

Joints of the foot

Before examining the muscles, study the joints of the foot, making comparisons with the hand.

The synovial **subtalar joint** (**7.6.3**, **7.6.4**) is the articulation beneath the talus. It consists of two parts, the articular surfaces of which are reciprocally curved so that they permit inversion and eversion of the foot. Study the lateral radiograph (see **7.2.26**) and sagittal MRI (**7.6.4**) of the ankle and subtalar joints. The posterior articulation (to which the term subtalar joint is often restricted) is between the concave facet on the under surface of the talus and the convex facet on the upper surface of the calcaneus (talo-calcaneal joint). Its fibrous capsule is strengthened by **medial** and **lateral ligaments** and by an **interosseous ligament** which occupies the bony **tarsal canal** between

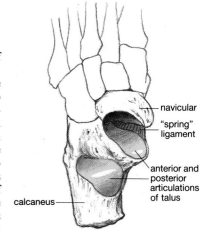

7.6.2
Enlarged and deformed 1st metatarso-phalangeal joint ('bunion').

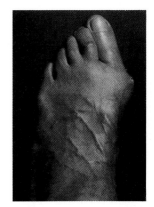

navicular

"spring" ligament

anterior and posterior articulations of talus

calcaneus

7.6.3
Subtalar joint.

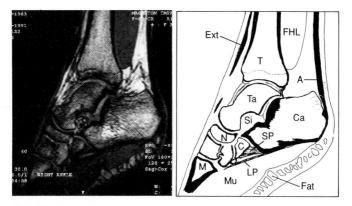

7.6.4
MRI of ankle, subtalar, and midtarsal joints
(sagittal section). T—tibia; Ta—talus;
Ca—calcaneus; N—navicular; C—cuboid;
M—metatarsal; Ext—extensor tendon;
FHL—flexor hallucis longus; A—Achilles
tendon; Si—tarsal sinus containing
interosseous ligament; SP—short plantar
ligament; LP—long plantar ligament;
Mu—muscles of sole; Fat—loculated
subcutaneous fat in sole.

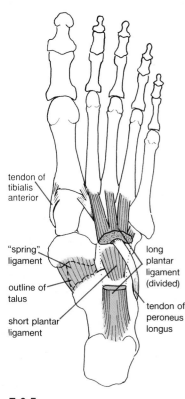

tendon of
tibialis
anterior

"spring"
ligament

outline of
talus

short plantar
ligament

long
plantar
ligament
(divided)

tendon of
peroneus
longus

7.6.5
Ligaments of sole of foot.

talus and calcaneus. The anterior articulation (talo-
calcaneo-navicular joint) is effectively a ball and
socket joint between the head of the talus and a
socket formed by the concave posterior surface of
the navicular, the anterior facet on the upper
surface of the calcaneus, and the upper surface of
the 'spring' ligament. The **'spring' (plantar
calcaneo-navicular) ligament (7.6.5)** is a
broad, thick band which connects the anterior
margin of the sustentaculum tali to the navicular. It
ties together the apex of the medial arch of the foot,
preventing the talus from being forced downward
between the calcaneum and navicular. Its upper
surface is usually covered wtih articular hyaline
cartilage forming part of the articular surface of the
talo-calcaneo–navicular joint. The calcaneum and
navicular are also connected on the dorsum of the
foot by a **bifurcated ligament** which, in addition, connects the calcaneum to the cuboid.

The **calcaneo-cuboid joint**, at the apex of the
'lateral arch' of the foot, has a capsule strengthened
dorsally by the bifurcated ligament and, on its
plantar aspect, by the short and long plantar
ligaments. With the calcaneo-navicular articulation it forms a **'midtarsal'** joint which passes
transversely across the foot and moves a little
during inversion/eversion movements. Minor
twisting injuries can cause dislocation of the mid-
tarsal and tarso-metatarsal joints which is difficult
to diagnose and causes prolonged discomfort.

The remaining inter-tarsal joints, tarso-metatarsal, metatarso-phalangeal, and interphalangeal
joints are very similar to their counterparts in the
hand. The joints of the great toe, like its bones, are
particularly strong, and the 1st metatarsal head is
grooved on its plantar surface for two sesamoid
bones which protect the flexor hallucis longus
muscle as it passes beneath the joint towards the
distal phalanx.

The **short plantar ligament (7.6.4, 7.6.5)** fills
the depression between the anterior tubercle of the
calcaneum and the ridge on the cuboid. The **long
plantar ligament (7.6.4, 7.6.5)** extends from a
broad attachment on the under surface of the
calcaneum to the ridge on the cuboid and forward
to the bases of the 2nd to 5th metatarsals.

Fascia and muscles of the sole of the foot

The subcutaneous **fat** of the sole and heel (**7.6.4**) is
thick and loculated by fibrous septa which pass
through it from the skin to the plantar aponeurosis.

Qu. 6E *What is the functional importance of this
arrangement?*

The **plantar aponeurosis** lying deep to the
subcutaneous tissue is very thick and strong; it
consists of both longitudinal and transverse bands
of fibres. It is attached posteriorly to the calcaneus;
anteriorly it divides into five slips which extend
forward into the toes to attach to the sides of the
proximal phalanges.

Examine the muscles and tendons of the four
layers of the sole of the foot. (Details of their origins
and insertions need not be memorized.)

Layer 1. (**7.6.6**) Beneath the plantar aponeurosis
abductor hallucis arises from the medial tubercle of the calcaneus and the surrounding fascia and
inserts into the medial side of the base of the
proximal phalanx of the big toe; it abducts and
flexes the big toe. **Abductor digiti minimi** arises
from both tubercles of the calcaneus and inserts on
the lateral aspect of the base of the proximal
phalanx of the little toe. **Flexor digitorum**

brevis arises from the medial tubercle of the calcaneus and the aponeurosis; its four tendons split to allow the passage of the tendons of flexor digitorum longus and then insert into the sides of the middle phalanges. As in the hand, the tendons lie in a fibrous and synovial sheath in the toes.

Qu. 6F *Which tendons in the hand are analogous to those of flexor digitorum brevis?*

Layer 2. (7.6.7) Beneath the muscles of Layer 1 lie the tendons of two of the deep calf muscles. The **flexor digitorum longus tendon** divides into four slips which pass through a hiatus in the tendons of flexor digitorum brevis to insert into the base of the terminal phalanx of the four lateral toes; the **flexor hallucis longus tendon** grooves the back of the talus and the sustentaculum tali, passes along the medial side of the sole to the 1st metatarso-phalangeal joint where it is protected in a groove formed by the two sesamoid bones in flexor hallucis brevis, and inserts into the base of the distal phalanx of the big toe. The 2nd layer of the sole of the foot also contains **flexor accessorius** which arises from the medial concave surface of the calcaneus and inserts into the deep surfaces of the tendons of flexor digitorum longus. Flexor accessorius acts to convert the oblique pull of the long tendons into a direct antero-posterior pull. As in the hand, four **lumbrical** muscles arise from the tendons of flexor digitorum longus, cross the metatarso-phalangeal joints on the tibial side, and insert into the dorsal extensor expansions, thus flexing the metatarso-phalangeal joints.

Qu. 6G *Grip the floor with your toes; how important is this action?*

Layer 3 (7.6.8) comprises three muscles which all act as their name implies: **flexor hallucis brevis**, which originates from the region around the spring ligament, has two bellies which insert into either side of the base of the proximal phalanx of the big toe. Sesamoid bones within these insertions prevent the tendons (and that of flexor hallucis longus) from being crushed by the weight of the body. **Adductor hallucis** has a transverse and an oblique head, the common tendon of which inserts with the lateral part of flexor hallucis brevis into the lateral sesamoid and side of the base of the proximal phalanx of the big toe. **Flexor digiti minimi brevis** arises from the plantar surface of the base of the 5th metatarsal bone and inserts into the lateral side of the base of the proximal phalanx of the little toe.

Layer 4 (7.6.9, 7.6.10) comprises the **plantar** and **dorsal interossei** which are inserted into the bases of the proximal phalanges and not into the dorsal extensor expansions as in the hand. They are organized around a functional axis running longitudinally through the 2nd toe. The flexion of the proximal phalanges that they produce is important for the grip of the toes, but abduction and adduction are little used when shoes are worn. Layer 4 also contains two tendons derived from muscles of the leg. The **tendon of tibialis posterior** passes behind the medial malleolus to insert into the tuberosity of the navicular bone; the **tendon of peroneus longus** passes obliquely across the sole in the groove on the cuboid and

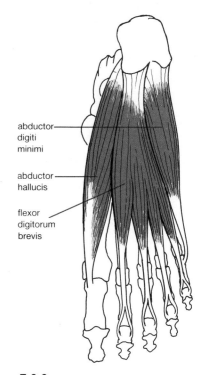

7.6.6
Sole of foot: muscle layer 1.

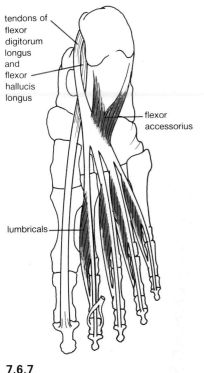

7.6.7
Sole of foot: muscle layer 2.

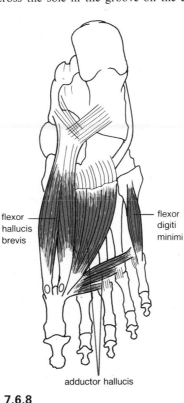

7.6.8
Sole of foot: muscle layer 3.

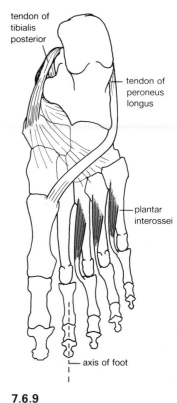

7.6.9
Plantar interossei: muscle layer 4.

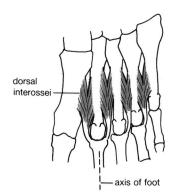

7.6.10
Dorsal interossei: muscle layer 4.

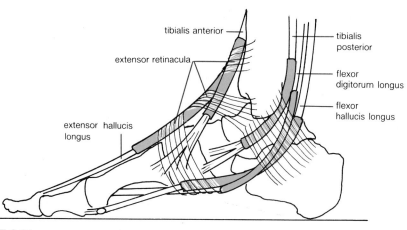

tibialis anterior

extensor retinacula

extensor hallucis longus

tibialis posterior

flexor digitorum longus

flexor hallucis longus

7.6.11
Extensor and flexor synovial tendon sheaths and retinacula.

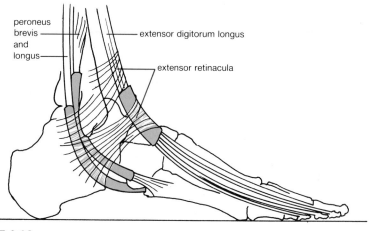

peroneus brevis and longus

extensor digitorum longus

extensor retinacula

7.6.12
Peroneal tendon sheaths and retinacula.

7.6.13
Development of footprints from 6 months to adult.

inserts into the medial cuneiform and the base of the 1st metatarsal. The **medial plantar branch of the tibial nerve** supplies abductor hallucis, flexor digitorum brevis, flexor hallucis brevis, and the first lumbrical. The **lateral plantar nerve** supplies the remaining muscles.

Review the muscles of the hand and compare them with those of the sole.

Qu. 6H *Which nerve is comparable to the median nerve in the hand?*

The retinacula and synovial tendon sheaths surrounding tendons at the ankle are illustrated in **7.6.11, 7.6.12**. Compare this arrangement with that at the wrist (see **6.7.12, 6.7.13**).

The arch of the foot (7.6.1)

The arch of the foot enables the weight of the body to be distributed between the medial and lateral tubercles of the calcaneus posteriorly and the heads of the lateral four metatarsals and two sesamoid bones of the first metatarsal anteriorly. The basic structure of the arch, which consists of a number of small bones united by many plane synovial joints, allows for considerable flexibility and spring, and this in turn facilitates walking and running. It also allows the foot to adapt in shape when walking on uneven ground.

The toes provide leverage for propelling the body forwards and, by their gripping action, help to provide stability and balance both in standing and during movement.

The arch of the foot is often analysed in terms of its longitudinal (medial and lateral) and transverse components, but the transverse component is only half an arch, as becomes apparent when the feet are placed together. These all consist of individual bones united by joints and reinforced by dorsal and plantar ligaments. The ligaments provide passive support for the arch; active support, which is particularly important during movement, is provided by muscles. These include the **small muscles of the sole of the foot** and the **long tendons of calf muscles**, in particular those of tibialis anterior, flexor hallucis longus, and flexor digitorum longus. Tibialis anterior pulls upward on the medial cuneiform and the base of the first metatarsal at the apex of the arch, while the long flexors 'tie' the extremities of the arch together. The **long** and **short plantar ligaments (7.6.5)** which extend, respectively, from the calcaneus to the cuboid and to the distal tarsal bones strengthen the lateral aspect of the arch and also form a fibrous tunnel for the tendon of peroneus longus. Medially the arch is subject to body weight transmitted through the tibia which will tend to push the head of the talus between the calcaneus and navicular bones. The strong **spring ligament (7.6.5)**, which extends between the sustentaculum tali and the navicular, supports the head of the talus and thereby maintains the arch. When standing passively, there is little activity in the muscles of the foot, so that the ligaments are the most important supports of the arch. Immediately the foot becomes active, muscles take the strain off the ligaments.

A baby's foot does not appear to possess the medial longitudinal arch. When the child begins to walk, towards the end of the first year or beginning of the second, the relatively large amount of fat regresses and the arch begins to appear **(7.6.13)**. This 'flat' appearance may, however, persist through later life and be the cause of chronic discomfort.

Requirements:

Articulated skeleton and separate foot bones
Prosections of the joints and ligaments of the foot; and of the four layers of the sole of the foot
Radiographs of the foot in different positions.

Seminar 7

Blood supply and lymphatics of the lower limb

Aims To study the arterial supply and venous drainage of the lower limb; to consider the vascular anastomoses; to consider adaptations of the vasculature associated with the upright stance; and to consider the course of lymphatic vessels and the locations of lymph glands in the lower limb.

A. Living anatomy (see 7.7.3)

It is important to be able to feel arterial pulses in the normal living leg because their absence in patients may signify arterial disease. Ask your partner to lie on a couch and feel first for the pulsation of the **femoral artery** in the groin. This is usually easy

to find. Press gently with one, or at most two fingertips just below the 'mid-inguinal point' (midway between the pubic symphysis and the anterior superior iliac spine) (7.7.1). Next, ask your partner to flex his knee to a right angle and press your fingertips into the popliteal fossa, toward the centre of the knee joint. You may be able to feel the pulsation of the **popliteal artery**, though this can be difficult. It may be easier to feel the artery with your partner lying prone.

Feel for the **dorsalis pedis artery** at the mid point between the medial and lateral malleoli. It continues on the dorsum of the foot lateral to the tendon of extensor hallucis longus until it reaches the proximal end of the first intermetatarsal space where it may also be palpable, before it passes through the space into the sole. The **posterior tibial artery** should easily be felt behind the medial malleolus, where it lies deep to the flexor retinaculum. The leg also receives blood supply via the gluteal arteries in the buttock, but these are deep and their pulsations cannot be felt.

Ask your partner to stand up and identify the **venous arch** on the dorsum of the foot. Trace it laterally to a **short saphenous vein** which runs behind the lateral malleolus, and medially to a **long saphenous vein** which runs in front of the medial malleolus, and up the medial aspect of the calf and thigh (see 7.7.9). You may be able to see the small normal dilatations of the veins associated with valves. Now ask your partner to lie down on a couch and note how the veins empty as the hydrostatic column of blood is reduced.

Qu. 7A *What problems has the adoption of the upright stance created for venous drainage of the lower limb?*

The ankle is a suitable site for intravenous infusion when arm veins are difficult to find. The long saphenous vein can usually be located if a transverse incision is made 2 cm above and 2 cm anterior to the medial malleolus.

Most of the lymphatic system cannot be seen or felt, but palpate your partner's groin just distal to the mid-part of the inguinal ligament, where you may be able to feel a few **superficial inguinal lymph nodes**.

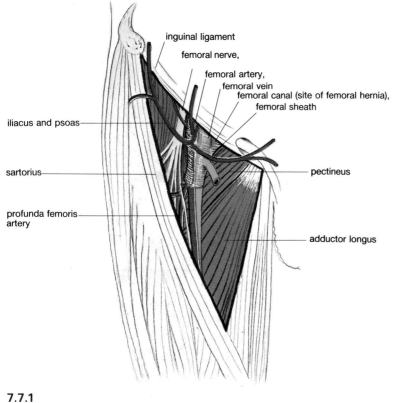

inguinal ligament
femoral nerve,
femoral artery,
femoral vein
femoral canal (site of femoral hernia),
femoral sheath
iliacus and psoas
sartorius
pectineus
profunda femoris artery
adductor longus

7.7.1
Femoral triangle.

B. Prosections

Arterial supply of the lower limb

When the **external iliac artery (7.7.2, 7.7.3)** passes under the mid-inguinal point to enter the thigh, it is renamed the **femoral artery**. In the thigh it lies first in the **femoral triangle (7.7.1)** which is bounded by the inguinal ligament above and the *medial* borders of sartorius and adductor longus below; adductor longus (medially), and pectineus and iliopsoas (laterally) form its floor. The femoral artery is accompanied on its medial side by the **femoral vein** and both lie in separate fascial compartments within the **femoral sheath**, a funnel-like prolongation beneath the inguinal ligament of fascia covering iliacus and fascia lining the anterior abdominal wall. In the femoral triangle the femoral artery gives a large branch, the **profunda femoris artery**, which arises from the back of the femoral artery and then runs distally passing deeply between adductor longus and adductor magnus. The profunda femoris provides the principal supply to the thigh. It gives off **medial** and **lateral circumflex femoral arteries**, which not only encircle the femur but also give ascending and descending branches which supply thigh muscles and the hip joint (via the trochanteric anastomosis), and **perforating arteries** which pierce the adductor muscles to supply the muscles of the adductor and hamstring compartments and the skin of the thigh.

Follow the femoral artery as it leaves the femoral triangle and passes distally into the subsartorial (adductor) canal (which lies between vastus medialis and the adductors and deep to sartorius). At the distal end of the canal the artery passes through an arch between the adductor and hamstring parts of adductor magnus to reach the popliteal fossa at the back of the knee joint. Here, the artery is renamed the **popliteal artery (7.7.5)**. It gives branches which supply the knee joint, and others which form

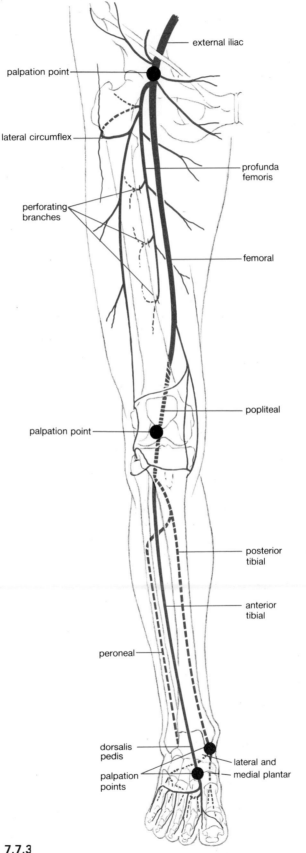

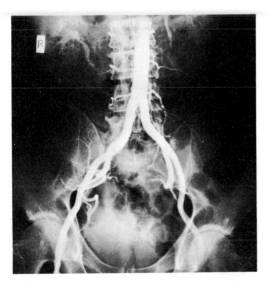

7.7.2
Arteriogram of iliac and femoral arteries.

7.7.3
Major arteries of lower limb: pressure points for the arrest of haemorrhage indicated ●●.

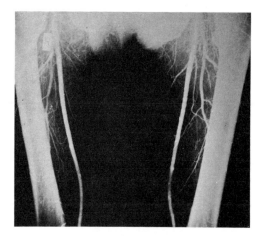

7.7.4
Arteriogram of femoral and profunda femoris arteries.

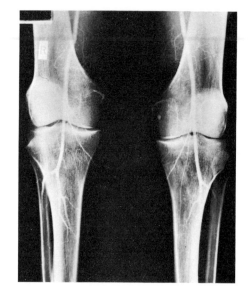

7.7.5
Arteriogram of popliteal artery.

anastomoses which encompass the lower end of the femur and the upper end of the tibia, linking the vessels above the knee with those distal to it.

After it leaves the popliteal fossa, the popliteal artery divides into **anterior** and **posterior tibial arteries**. Identify and follow the **anterior tibial artery** as it passes between the tibia and fibula above the interosseous membrane and descends on its anterior surface to supply the muscles of the anterior compartment of the leg and **medial** and **lateral malleolar branches** to the ankle. When it crosses the front of the ankle joint it is renamed the **dorsalis pedis artery** (**7.7.6**); this passes forward over the dorsum of the foot to reach the interval between the 1st and 2nd metatarsal bones through which it passes to reach the sole of the foot and anastomose with the plantar arterial arch. Before it passes into the sole it gives a **transverse metatarsal** branch which runs laterally to supply small paired branches to the toes.

Now return to the **posterior tibial artery** which supplies the superficial and deep muscles of the calf and passes with the long flexor tendons around the back of the medial malleolus. It gives off a **peroneal artery** which runs distally deep to flexor hallucis longus (it can pass through to the dorsum of the foot and take over the distribution of the dorsalis pedis artery). At the ankle, the posterior tibial artery passes deep to the flexor retinaculum where it divides into medial and lateral plantar arteries which enter the sole. The **medial plantar artery** is small and runs along the medial side of the foot. The **lateral plantar artery** runs obliquely across the sole of the foot between the 1st and 2nd layers of muscles toward the base of the 5th metatarsal. The artery then passes deeply between the 3rd and 4th layers and runs medially across the interossei forming a **plantar arterial arch**, which gives off **metatarsal branches** and ends by anastomosing with the dorsalis pedis artery (which has reached the sole from the dorsum of the foot).

Qu. 7B *How comparable is this arrangement to that found in the palm of the hand?*

Qu. 7C *If a patient who complained of intermittent pain in the legs (intermittent claudication) during and after exercise also had a toe which was becoming gangrenous, what would you deduce? How might such a patient be treated?*

Look again at a prosection of the anterior compartment of the leg. The muscles it contains (tibialis anterior, extensor hallucis longus, extensor digitorum longus, and peroneus tertius) are enclosed by firm inextensible walls which comprise the deep fascia of the leg, the tibia, the interosseous membrane, the fibula, and the fascia separating the anterior compartment from the peroneal compartment. Compartmentation of muscle in the leg is advantageous to peripheral venous return because

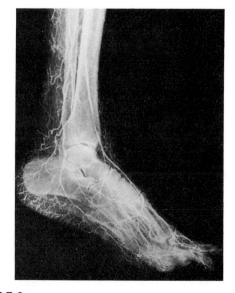

7.7.6
Arteriogram of dorsalis pedis and transverse metatarsal arteries.

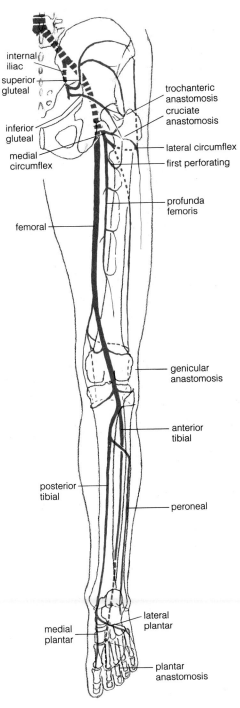

internal iliac
superior gluteal
inferior gluteal
medial circumflex
femoral
posterior tibial
medial plantar

trochanteric anastomosis
cruciate anastomosis
lateral circumflex
first perforating
profunda femoris
genicular anastomosis
anterior tibial
peroneal
lateral plantar
plantar anastomosis

7.7.7
Anastomoses between arteries of lower limb; posterior view.

it forms a muscular pump which forces blood centripetally, against gravity. However, if unaccustomed severe and prolonged exercise is undertaken by, for instance, an army recruit, the muscles of the anterior compartment may swell from fatigue. The swelling may be such that, in the enclosed space, the venous return is obstructed; this can lead to further swelling, and eventually to capillary stasis and gangrene. The condition is termed the 'anterior compartment syndrome' of the leg and if it is not relieved the results can be crippling. Provided that the problem is diagnosed early enough, it can be relieved by dividing the deep fascia along the whole length of the compartment.

Qu. 7D *Could the same problem develop with the muscles of the calf?*

The **superior** and **inferior gluteal arteries**, both branches of the internal iliac artery, supply the gluteal region. They also send small branches into the posterior compartment of the thigh where they make anastomoses with branches of the femoral (external iliac) system.

Anastomoses between arteries of the lower limb

A number of important anastomoses link the arteries of the lower limb to form a chain of anastomoses in the thigh (**7.7.7**). An anastomosis in the trochanteric fossa (**trochanteric anastomosis**), provides vessels which pass along the neck of the femur to supply the head of the femur. It is formed by branches of the gluteal arteries with ascending branches of the medial and lateral femoral circumflex arteries. A little more distally, at the junction between quadratus femoris and the upper border of adductor magnus, transverse branches of the circumflex arteries anastomose with a descending branch from the inferior gluteal artery and with the upper perforating branch of the profunda femoris to form the **cruciate anastomosis**. Still lower, the perforating arteries anastomose among themselves and with genicular vessels. At the knee, genicular branches, and at the ankle calcaneal branches of the principal arteries anastomose. A branch of the peroneal artery (**7.7.3**) which passes anteriorly between the tibia and fibula may be prominent if the dorsalis pedis artery is small. In the foot the dorsalis pedis branch of the anterior tibial artery anastomoses with the plantar arch derived from the posterior tibial artery. You should consider ways in which the leg can derive arterial supply if the femoral artery is blocked at different sites.

The blood vessels of the lower limb are innervated by sympathetic autonomic fibres. Preganglionic fibres originate from cell bodies situated in the grey matter of lower thoracic and highest lumbar segments of the spinal cord. Postganglionic fibres arise from the lumbar sympathetic chain and form a plexus around the external iliac artery. Other fibres from the lumbar and pelvic parts of the chain join the femoral, obturator, and sciatic nerves in order to reach the vessels and their branches. If the arterial flow in the lower limb is inadequate, the lumbar sympathetic nerves may be surgically divided in an attempt to reduce any vasoconstrictor sympathetic tone.

C. Radiographs

Study the arteriograms (**7.7.5**, **7.7.6**) and identify the various branches shown. After looking at the

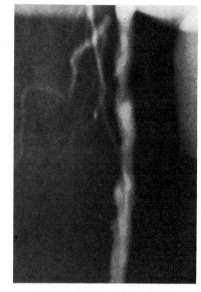

7.7.8
Arteriogram showing blockage due to
atheroma.

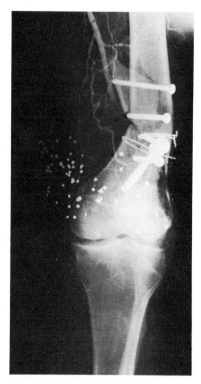

7.7.9
Arteriogram of gun-shot wound to knee;
blocked popliteal artery.

normal femoral arteriogram, look at **7.7.8**. The
difference in appearance is due to the presence of
atheroma, a cholesterol-containing deposit in the
tunica intima of the arterial wall. This narrows the
lumen of the vessel, sometimes blocking it completely. Gangrene of the distal limb might occur in
such cases.

Study arteriogram **7.7.9**. What has happened?
The leg suffered a gunshot injury, shattering the
bone and destroying the popliteal artery. The
fractured bones were stabilized as well as was
possible, but gangrene of the foot would have
occurred unless the blood supply had been restored
surgically.

Venous drainage of the lower limb (7.7.10)

As in the upper limb there is a system of **superficial veins** which drain the skin and superficial
fascia, and a system of **deep veins** which accompany the arteries, commencing as small, paired
venae comitantes. These two systems, which are
separated by the deep fascia which envelops the
muscle compartments, are connected by a number
of very important **communicating veins**.

A. Prosections

Superficial venous drainage

On a superficial prosection of the dorsum of the
foot locate the dorsal venous arch which drains the
toes and veins which pass dorsally from the sole.
The lateral aspect of the arch is drained by the
small (short) saphenous vein which runs
behind the lateral malleolus and up the back of the
calf to pierce the deep fascia of the popliteal fossa
and drain into the popliteal vein (**7.7.10**). The
medial side of the dorsal venous arch is drained by
the larger **great (long) saphenous vein** which
passes in front of the medial malleolus where it can
usually be seen and felt. This major vein then
ascends unsupported in the subcutaneous tissue
along the medial aspect of the leg and thigh. At the
top of the thigh, about 5 cm below the pubic
tubercle, it enters a saphenous opening in the fascia
lata and empties into the **femoral vein** in the
femoral triangle. At this point several smaller veins
draining the iliac and pudendal regions also join the
femoral vein, and it is here that the great saphenous
vein can conveniently be tied off if it becomes
varicose (**7.7.12**). Open the great saphenous vein
and examine it for the presence of valves which
divide up the column of blood and thus reduce the
hydrostatic pressure.

Deep drainage

Two venae comitantes accompany each of the
smaller arteries in the leg and eventually join to
form the **popliteal vein**. This vessel lies in the
popliteal fossa superficial to the popliteal artery.
The popliteal vein passes with the artery through
the hiatus in adductor magnus to enter the
adductor compartment of the thigh, where it is
renamed the femoral vein. The femoral vein runs
proximally through the adductor canal to the
femoral triangle, receiving many large tributaries
including the profunda femoris vein and the long

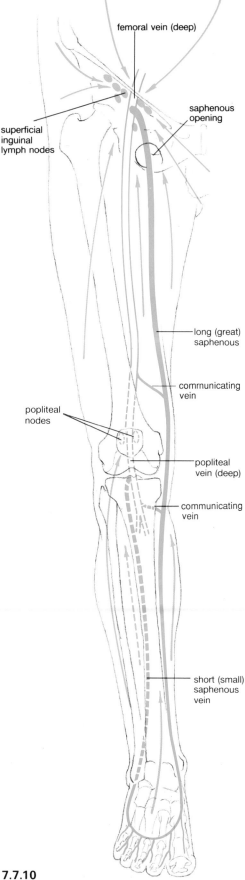

7.7.10
Major veins of the lower limb; lymphatic
drainage of lower limb indicated.

saphenous vein. It then passes deep to the inguinal
ligament to enter the pelvis where it is renamed the
external iliac vein. When you stand (and under
various other circumstances), the venous pressure
in the lower limb increases and the femoral vein
distends into the femoral canal, a potential space on
the medial side of the femoral vein within the
femoral sheath; the femoral canal also transmits
lymphatics from the leg and external genitalia to
the abdomen. The femoral canal is a continuation
of the abdominal cavity (not the peritoneal cavity)
so that occasionally, when intra-abdominal pressure is increased, abdominal contents can follow
the path of least resistance to cause a femoral
hernia.

Within soleus are many deep veins which are
difficult to demonstrate in the cadaver. The blood
in these veins is especially liable to clot (thrombose)
in circumstances of injury or disease (particularly
post-operatively) in which immobilization and
stasis of flow occurs. The presence of such a deep
vein thrombosis can be demonstrated by means of a
venogram. Radio-opaque dye is injected into a
superficial vein of the foot and a radiograph is
taken shortly afterwards. **7.7.11** shows a normal
venogram.

Communicating veins

A system of communicating veins joins the superficial veins, especially the great saphenous vein, to
the deep veins running with the major arteries.
These veins necessarily pierce the fascia lata. If they
are dilated, the point at which they pierce the fascia
can be palpated as a tender defect which admits the
tip of a little finger. Major communicating veins are
found just above and just below the knee joint, on
its medial aspect. Search for these veins on a
prosected specimen. The valves in communicating
veins normally permit blood to flow from the
superficial to the deep veins but not vice versa.

Venous return and the 'muscle pump'

The contraction of muscles within a relatively
constricted fascial envelope, together with the
venous valves, provides a 'muscle pump' which
actively empties the deep veins of blood, forcing it
proximally within the veins. As the deep veins are
emptied and the muscle relaxes, the valves in the
communicating veins allow blood in the superficial
veins to drain into the deep veins, from which it is
then pumped. Thus the muscle pump empties the
deep veins directly and the superficial veins indirectly. If, for congenital or other reasons, the venous
valves become incompetent, the hydrostatic columns of blood become greater and the superficial
veins become progressively more dilated, leading to
what are termed **varicose veins (7.7.12)**. If the
valves in communicating veins become incompetent the action of the muscle pump only exacerbates
the problem, because blood is now pumped *into* the
superficial veins rather than away from them. The
drainage of the skin by superficial veins becomes
compromised, leading to skin changes which can
progress to ulceration.

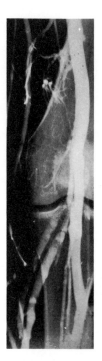

7.7.11
Normal venogram; short
saphenous vessels entering
popliteal vein.

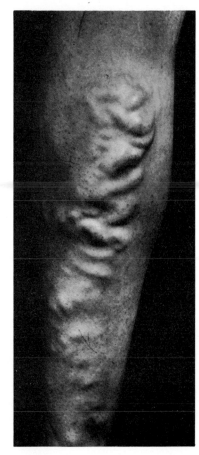

7.7.12
Varicose veins on calf.

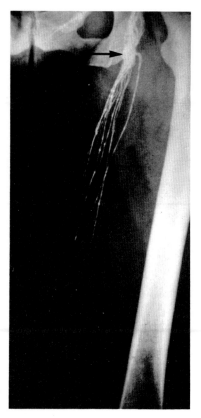

7.7.13
Lymphangiogram of vessels
draining into superficial inguinal
lymph nodes.

Lymphatic drainage

(7.7.10, 7.7.13)

The lymphatic drainage of the lower limb follows the general pattern found in the upper limb: superficial lymphatics running with superficial veins drain the skin and subcutaneous tissue while deep lymphatics running with the neurovascular bundles drain structures deep to the deep fascia. However, in the lower limb there is very little communication between superficial and deep lymphatics. Superficial lymph nodes (arrow; **7.7.13**) are found in the superficial fascia just below the inguinal ligament and at the entry of the long saphenous vein into the femoral vein. These superficial inguinal nodes drain the skin and superficial fascia of the lower limb and of the trunk below the waist line. Since the lower part of the anal canal, the lower part of the vagina, and the external genitalia develop from ectoderm, these also drain into the superficial inguinal lymph nodes. A group of deep inguinal nodes around the femoral artery drains the deep tissues.

The lymph vessels leaving the superficial inguinal nodes pierce the fascia lata around the saphenous opening to drain into the deep inguinal nodes. Their efferent lymphatics pass upward through the femoral canal to join the chain of nodes that lie along the external and common iliac arteries and aorta. By this route they reach a distended sac, the cisterna chyli, which lies between the aorta and the bodies of the L1 and L2 vertebrae. This cistern is drained by the thoracic duct which passes upward on the posterior wall of the thorax and eventually joins with lymphatics from the head and the left upper limb (subclavian trunk, p. 72) to drain into the confluence of the left subclavian and internal jugular veins. In this way, the lymph is returned to the blood stream.

Requirements:

Articulated skeleton of lower limb
Prosections of the arterial supply to the lower limb showing femoral triangle and passage of femoral artery through adductor magnus to popliteal fossa; posterior tibial arteries, medial and lateral plantar arteries; dorsalis pedis artery; gluteal arteries
Prosections of the superficial veins and any lymph nodes
Arteriograms, venograms, and lymphograms of the lower limb.

Seminar 8

Innervation of the lower limb: lumbosacral plexus, femoral and obturator nerves

Aims To consider the principles which underlie the innervation of the lower limb; to study the lumbo-sacral plexus by which motor and sensory nerves are distributed to muscle groups, joints, and skin of the lower limb; to study the distribution of the femoral and obturator nerves to the muscles and skin of the anterior (extensor) and medial (adductor) compartments of the thigh respectively.

The innervation of the lower limb is derived from the lumbo-sacral plexus which arises from the roots of lumbar spinal nerves deep to psoas within the abdomen and from the roots of sacral spinal nerves on the posterior wall of the pelvis (**7.8.1**).

The arrangement of the segmental nerve supply to the muscles and skin has essentially the same pattern as in the upper limb. Movements at most joints are controlled by four adjacent segments of the spinal cord; the upper two segments innervate one movement, the lower two the opposing move-ment. More distal muscles are supplied by more distal segments of the spinal cord (Table 2).

As in the upper limb, progressively more caudal nerve roots supply the skin of the preaxial border, the distal extremity, and the postaxial border of the lower limb (**7.8.2**). There is considerable overlap of supply between adjacent dermatomes, but less across axial lines, which are less well defined than in the upper limb. (S4 and S5 dermatomes supply the perineum.)

Postganglionic sympathetic fibres to the lower limb are derived from cells of the lumbar and sacral sympathetic ganglia. They either join the spinal nerves and are distributed with their branches, or form a plexus which passes distally along the arteries. They supply blood vessels and sweat glands.

A. Living anatomy

Following Table 2, make the movements of your lower limb joints and note the nerve roots which 'supply' each movement.

With a skin pencil mark out the dermatome lines on your partner. The dermatome pattern of the lower limb is considerably distorted compared with that of the upper limb, largely because the lower limb has become medially rotated during develop-ment. As in the upper limb the middle spinal nerve root of the plexus supplies the most distal skin (i.e. the sole of the foot is supplied mainly by L5).

Qu. 8A *Which spinal nerve supplies the area of skin on which you sit?*

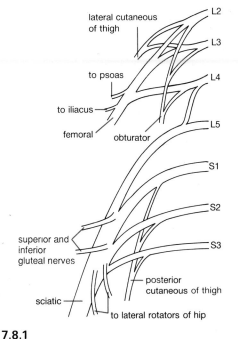

7.8.1
Lumbosacral plexus.

Table 2 Segmental nerve supply to movements of the lower limb

Joint	Muscle action	Root supply	Muscle action	Root supply
Hip	Flexors adductors med. rotators	L1,2,3	Extensors abductors lat. rotators	L4,5, S1
Knee	Extensors	L 3,4	Flexors	L 5, S1
Ankle	Dorsi–flexors	L 4,5	Plantar–flexors	S1, 2
Subtalar	Invertors	L 4,5	Evertors	L 5, S1
Foot	Toe extensors	L 5, S1	Toe long flexors	S 2
			Small muscles of sole	S3

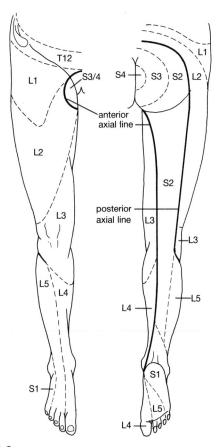

7.8.2
Dermatomes of lower limb; note the axial lines.

B. Dissection and prosections

Femoral nerve (7.8.3, see 7.7.1)

On a prosection identify the lumbar plexus deep to psoas major and the **femoral nerve** (L2,3,4; posterior divisions) which emerges from its lateral border and runs distally in the groove between psoas and iliacus to pass beneath the inguinal ligament. If you have not removed the skin from the lower limb before reaching this seminar, do so now. Incise and remove the thick fascia lata which ensheathes the thigh, covering the femoral triangle. As you do so, note cutaneous nerves which pierce the fascia. Identify the inguinal ligament, sartorius, adductor longus, and the femoral artery and vein which lie in the femoral triangle within the femoral sheath. Now locate the femoral nerve as it emerges from beneath the inguinal ligament outside and lateral to the femoral sheath where it breaks up into its terminal branches. Trace these branches and locate **muscular branches (7.8.3)** to all the muscles of the anterior compartment of the thigh.

Qu. 8B *When you elicit a 'knee jerk', what elements of this stretch reflex are you testing?*

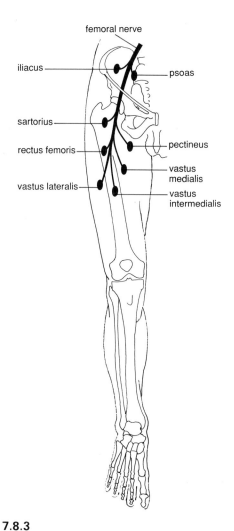

7.8.3
Femoral nerve; supply to muscles.

Sensory branches of the femoral nerve (**7.8.4**) supply the skin over the front and medial aspects of the thigh (**medial** and **intermediate cutaneous nerves of thigh**). Find these nerves and trace the **saphenous nerve**, a long sensory branch which runs through the adductor canal and then accompanies the long saphenous vein, supplying a strip of skin which runs distally from the medial aspect of the knee to the 1st metatarsophalangeal joint. To trace these branches you will find it easier to cut across the sartorius and reflect it. The femoral nerve also sends sensory branches to the hip and knee joint along the muscles which act on those joints.

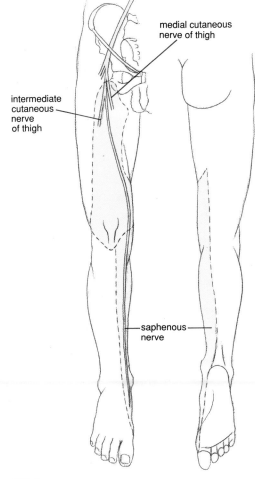

7.8.4
Femoral nerve; cutaneous distribution.

The **lateral cutaneous nerve of thigh** arises separately from the lumbar plexus, passes under the lateral end of the inguinal ligament, and supplies the skin of the lateral aspect of the thigh. It can become involved in a disorder of fibrous tissue which traps the nerve beneath the inguinal ligament, causing pain down the lateral side of the thigh.

Obturator nerve (7.8.5)

On a prosection of the pelvis locate the **obturator nerve** (L2,3,4; anterior divisions) as it emerges from the medial aspect of psoas major within the pelvis and runs distally toward the obturator foramen to pass through a canal in the upper part of the foramen and reach the thigh.

The obturator nerve supplies all the muscles of the adductor compartment (**7.8.5**) and sends sensory branches to the skin over it (**7.8.6**). It also supplies the hip and knee joints. Cut through adductor longus half-way along its length and locate adductor brevis deep to it. Identify the **anterior division of the obturator nerve** as it lies between adductors longus and brevis. Trace it proximally to the obturator externus muscle, which the obturator nerve supplies and pierces on leaving the pelvis. Deep to adductor brevis locate the **posterior division of the obturator nerve** lying on adductor magnus.

> **Qu. 8C** *Why might women feel monthly discomfort on the medial aspect of the thigh?*

The skin of the medial side of the thigh adjacent to the groin is supplied by the ilio-inguinal nerve (L1), a branch of the lumbar plexus which runs through the abdominal wall to supply the scrotum. A reflex which elevates the testes (cremasteric) can be elicited by stroking this area of skin (see Vol. 2). Also, the femoral branch of the genito-femoral nerve (L1,2) supplies a handsbreadth area just distal to the inguinal ligament, emerging from the abdomen superficial to the femoral artery.

Requirements:

Articulated skeleton
Prosections of the abdomen and pelvis showing the lumbosacral plexus and the femoral and obturator nerves
Prosections of anterior and medial compartments of the thigh showing the femoral and obturator nerves and their branches.

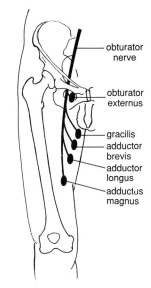

7.8.5
Obturator nerve; supply to muscles.

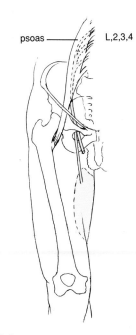

7.8.6
Obturator nerve; cutaneous distribution.

Seminar 9

Innervation of the lower limb: sciatic nerve

Aims To study the sciatic nerve and other branches of the sacral plexus that supply the lower limb; to study the distribution of the sciatic nerve to the posterior compartment of the thigh, and of its principal branches to the flexor and extensor muscles of the leg and foot; and to consider the disabilities which arise if either the nerves or their roots in the lumbosacral plexus are damaged.

A. Living anatomy

In the buttock and thigh the **sciatic nerve** lies deeply and cannot be felt. Its principal branch to the extensor muscles of the ankle and foot, the **common peroneal nerve**, can be felt as a distinct cord as it winds round the neck of the fibula before passing deeply into peroneus longus. The principal branch to the flexor muscles, the **tibial nerve**, is superficial only behind the medial malleolus, but even there it is difficult to feel.

B. Dissection and prosections

Branches of the sacral plexus—Sciatic nerve (see **7.8.1**)

On a prosection of the pelvis, locate the large sacral nerve roots as they emerge from the anterior sacral foramina and leave the pelvis above and below piriformis. Note that a branch nerve from L4 joins L5 to form the **lumbosacral trunk** which passes down from the lumbar plexus over the sacroiliac joint to join the sacral plexus.

Turn the cadaver face down. If you have not already done so when studying the gluteal region (see **7.3.7**), cut through gluteus maximus close to its origin and reflect it laterally. To do so you will also have to sever the muscle fibres which originate from the sacrotuberous ligament, and the **inferior gluteal nerve** (L5, S1,2) and artery which emerge from the pelvis below piriformis and pass directly into gluteus maximus. Identify gluteus medius and locate the **superior gluteal nerve** (L4,5, S1) and artery which emerge from the pelvis above piriformis to supply gluteus medius, gluteus minimus, and tensor fasciae latae. Locate the **posterior cutaneous nerve of the thigh** which lies superficial to the sciatic nerve and runs down the posterior

aspect of the thigh beneath the fascia lata supplying the area of skin that its name suggests.

The **sciatic nerve** (L4,5 S1,2,3) (**7.9.1**, **7.9.3**, **7.9.5**) supplies all the muscles of the posterior compartment of the thigh and, through its principal branches, all the muscles and most of the sensation below the knee. Identify this large nerve as it emerges from the pelvis below piriformis (see **7.3.7**); note that it is separated from the hip joint only by the small lateral rotator muscles of the hip. Trace it distally as it passes deep to biceps femoris to enter the posterior compartment of the thigh, where it

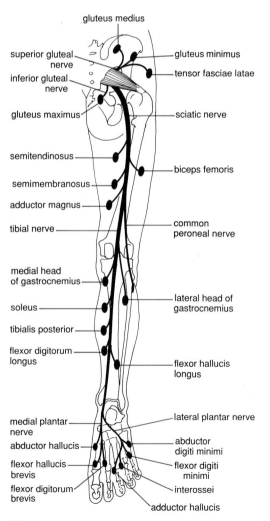

7.9.1
Sciatic and tibial nerves; supply to muscles.

supplies all the muscles including the hamstring part of adductor magnus. The sciatic nerve usually divides into its component **tibial** and **common peroneal nerves** half-way down the thigh, but the two bundles are distinct and may separate much higher in the thigh.

The **tibial nerve**, which is derived from anterior divisions of L4,5, S1,2,3, supplies all the flexor muscles of the back of the leg and the sole of the foot; the skin of the lower half of the back of the leg and the heel and sole of the foot and the joints of the knee, ankle, and foot. Trace the tibial nerve as it passes distally through the popliteal fossa, where it lies superficial to the popliteal vessels. It passes into the leg between the two heads of gastrocnemius and deep to the origin of soleus, supplying all three. Cut across the Achilles tendon and reflect these superficial muscles of the calf upward. Follow the tibial nerve from the popliteal fossa into the calf where it lies between the superficial and deep muscles and supplies all the deep muscles. Trace the nerve to the back of the medial malleolus and note its position with respect to the tendons and blood vessels, all of which are passing into the sole of the foot. Immediately behind the malleolus lies the tendon of tibialis posterior and the tendon of flexor digitorum longus, then the neurovascular bundle, and most posteriorly the tendon of flexor hallucis longus which grooves the talus as it enters the foot. Here the tibial nerve divides into its **medial** and **lateral plantar branches** (**7.9.4**) which behave in a similar fashion, respectively, to the median and ulnar nerves of the palm of the hand.

Now return to the sciatic nerve and trace its other terminal branch, the **common peroneal nerve** (**7.9.3, 7.9.7**) which is formed from posterior divisions of the nerve roots. It therefore supplies the extensor (including peroneal) muscles and much of the extensor skin of the leg and foot (**7.9.5, 7.9.6**).

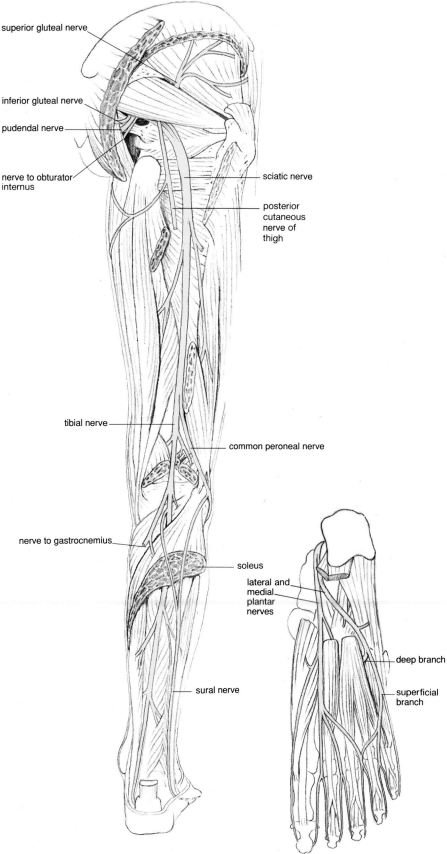

superior gluteal nerve

inferior gluteal nerve

pudendal nerve

nerve to obturator internus

sciatic nerve

posterior cutaneous nerve of thigh

tibial nerve

common peroneal nerve

nerve to gastrocnemius

soleus

lateral and medial plantar nerves

sural nerve

deep branch

superficial branch

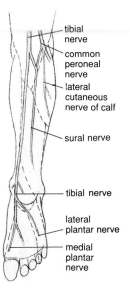

tibial nerve

common peroneal nerve

lateral cutaneous nerve of calf

sural nerve

tibial nerve

lateral plantar nerve

medial plantar nerve

7.9.2
Sciatic and tibial nerves; cutaneous distribution.

7.9.3
Course of sciatic and tibial nerves.

7.9.4
Course of medial and lateral plantar nerves.

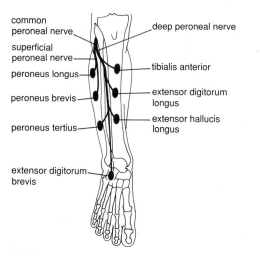

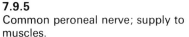

7.9.5
Common peroneal nerve; supply to muscles.

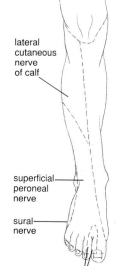

7.9.6
Common peroneal nerve; cutaneous distribution.

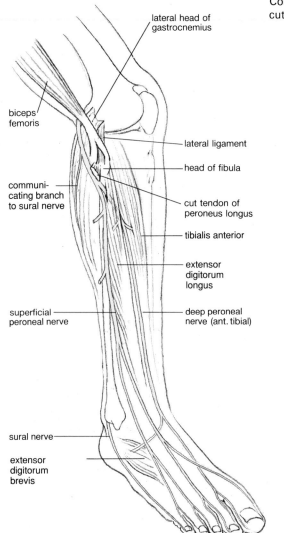

7.9.7
Course of superficial and deep peroneal nerves.

The common peroneal nerve passes through the popliteal fossa closely applied to the medial aspect of biceps femoris and then curves forward around the lateral aspect of the neck of the fibula. At this point it is superficial and therefore liable to injury either from direct trauma or from pressure caused by a badly applied plaster cast. The nerve supplies branches to the knee joint and branches (**lateral cutaneous nerve of calf**; part of the **sural nerve**) which supply skin on the back and posterolateral aspects of the calf. It then passes deep to peroneus longus and divides into the superficial and deep peroneal nerves.

Follow the **superficial peroneal nerve (7.9.7)** into the lateral (peroneal) compartment of the leg where it supplies peroneus longus and brevis. About two-thirds of the way down the lateral aspect of the leg the nerve emerges through the deep fascia and runs distally over the front of the leg and ankle. Here it divides into branches which supply the skin over the lower lateral aspect of the leg, the front of the ankle, and the dorsum of the foot (but not the cleft between the 1st and 2nd toes).

Now trace the **deep peroneal nerve (7.9.7)** which enters and supplies the anterior compartment of the leg. It descends on the interosseous membrane accompanied by the anterior tibial vessels which have joined it from the calf by passing over the interosseous membrane. The neurovascular bundle crosses the ankle midway between the two malleoli (i.e. between the tendons of extensor hallucis longus medially and extensor digitorum longus laterally) where it is vulnerable to sharp objects dropped on to the foot. In the foot it supplies extensor digitorum brevis and joints of the ankle and the foot. It ends by running forward to supply the skin of the cleft between the big and second toe.

The **sural nerve** is formed by branches of both the tibial and the common peroneal nerves. It runs down the back of the calf alongside the small saphenous vein and supplies the skin of the back of the calf, heel, and lateral side of the foot. This is the nerve most commonly used by surgeons for grafting a gap between the ends of an injured nerve.

Effects in the leg of nerve damage

The lumbosacral plexus is protected within the abdomen and the femoral, obturator, and sciatic nerves are unlikely to be damaged except by penetrating injuries. Loss of sympathetic fibres in damaged peripheral nerves causes dry skin with a 'blotchy' appearance, because regulation of sweat gland secretion and blood vessel calibre is lost.

Qu. 9A *A cyclist received a glancing blow on the side of the leg from a passing car which severed the common peroneal nerve. What would be the results of this injury?*

Qu. 9B *If a patient sustained a gunshot wound which severed a sciatic nerve in the gluteal region, what would be the results?*

Damage to lumbar and sacral nerve roots

Damage to individual or multiple spinal nerve roots can occur as a result of pressure from prolapsed intervertebral discs as they leave the vertebral canal through the intervertebral foramina (p. 148). The results can be predicted from a knowledge of the dermatomes and the segmental supply to muscles and it is often important to differentiate, by symptoms and clinical signs, which roots are involved. The most commonly affected roots are L4, L5 and S1. Disturbance of sensory roots can cause either pain or blunting of sensation in the area of its distribution; disturbance of motor roots causes weakness and ultimately wasting of the muscles innervated.

L4 root damage

Sensory signs. Sensation may be blunted over the antero-medial aspect of the shin and as far as the metatarso-phalangeal joint of the great toe (i.e. the distribution of the saphenous nerve).

Motor signs. It is not possible by clinical examination to detect wasting or weakness of all the muscles which are partially supplied by L4, but some wasting of quadriceps femoris, which receives a large contribution from L4, may be apparent. Tibialis anterior and tibialis posterior also receive a significant contribution from L4, so that weakness of foot inversion may be detectable.

Reflex jerks. Because of the effect on quadriceps the knee jerk may be reduced compared with the opposite side. The ankle jerk is not affected.

L5 root damage

Sensory signs. Blunting of sensation may extend down the anterolateral aspect of the shin across the dorsum of the foot to the hallux beyond the first metatarso-phalangeal joint; also the sole of the foot.

Motor signs. Slight weakness of extension (dorsiflexion) of the foot and extension of the big toe at the metatarso-phalangeal joint may be detected.

Extensor hallucis longus seems especially vulnerable to loss of function of this root. There may also be some weakness of eversion and wasting of the peronei.

Reflex jerks. Neither the knee jerk nor the ankle jerk are affected.

S1 root damage

Sensory signs. Blunting of sensation may extend down the posterolateral aspect of the calf to the dorsal and plantar surfaces of the outer aspect of the foot.

Motor signs. There may be loss of power of gluteus maximus and of (plantar) flexion of the ankle.

Reflex jerks. The knee jerk is unaffected, but the ankle jerk may be lost.

Referred pain in the leg

The hip joint is supplied by branches of the femoral, obturator, and superior gluteal nerves, and by the nerve to quadratus femoris. The root values of these nerves extend from L2 to S1. The knee joint is supplied by the femoral, obturator, tibial, and common peroneal nerves; these root values extend from L2 to S3. There is thus considerable overlap of the sensory supply of the two joints. This influences the presentation of disease: patients commonly complain of pain down the front of the thigh and knee even though their disease lies in the hip joint. For instance, early slipping of the upper femoral epiphysis often causes discomfort in the knee (see p. 96). Also, disease within the pelvis which affects the sciatic nerve can cause pain referred down the back of the thigh.

Requirements:

Articulated skeleton
Prosections of the pelvis showing the sacral plexus
Prosections of the sciatic nerve and its branches; common peroneal, superficial and deep peroneal, tibial, medial and lateral plantar nerves.

CHAPTER 8

The spinal column

Bipedalism in hominids emerged about two million years ago as a result of a combination of selection pressures concerned with object carrying, use of tools and weapons, and the need for vision over an increased distance. When *Homo sapiens* appeared the multiunit flexible spine had evolved long before, with the development of specialized joints situated between the spine and skull which enabled the head to rotate, thereby increasing the horizontal range of visual scanning. In addition, a series of alternating curvatures of the spine (forward-facing in the neck and lumbar regions, and backward-facing in the chest and pelvis) also appeared, ensuring the efficient balance of the head on the neck and of the trunk on the lower limbs, and supporting the thoracic cage and shoulder girdle. The erect posture exposed more of the vulnerable abdomen. This disadvantage was minimized by a reduction in the number of lumbar vertebrae, which brought the ribs closer to the pelvis and thereby increased protection. Maintenance of the erect posture also meant that the weight of a large part of the body had to be transmitted through the lumbar spine and sacrum to the pelvis and lower limbs, the forces involved being magnified many times in running and jumping. Although the spinal column has become well adapted for these functions in many ways, nevertheless, spinal problems are relatively common in modern man, occurring more frequently in the lower lumbar spine than in any other region.

Aims To study the type and extent of movement possible in different regions of the spine; to study the components of the multi-unit vertebral column, their regional differences, and adaptations for weight-bearing in the upright stance; to study the articulations and the muscles of the spine in relation to the movements which occur; to consider the protection provided by the spine for the spinal cord and its nerve roots.

The spine forms the central axis of the body and, by its shape, dominates the appearance of the trunk. It consists of a series of vertebrae, ligaments, and intervertebral discs. This flexible column of bone and ligament encloses a canal and protects the spinal cord within it. It supports the weight of the trunk and transfers the weight to the pelvis; the discs provide shock-absorbing resilience. It can be moved by a series of postural and 'prime mover' muscles to orient the limb girdles and the head; it also provides attachment for many other muscles (e.g. girdle muscles) and for the ribs. The vertebral

bodies have an additional function since they contain red bone marrow throughout life.

A. Living anatomy

Compare your partner's back (**8.1**), with the back of an articulated skeleton (**8.2**) and with radiographs of appropriate regions of the spine. Note that the spinal column normally lies strictly in the median sagittal plane (provided that the legs are of equal length).

Stand to one side and note the three spinal curvatures: **cervical lordosis** (dorsalward concavity); **thoracic kyphosis** (dorsalward convexity); and **lumbar lordosis**. Compare the spinal curvature of different subjects, particularly the degree of thoracic kyphosis. One way to assess this roughly and quickly is to ask the subject to stand

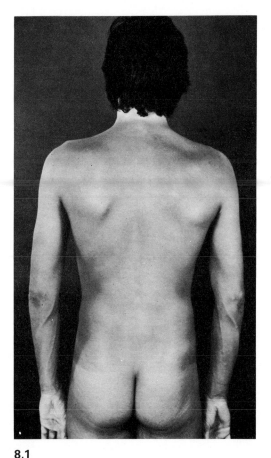

8.1
Normal back (male).

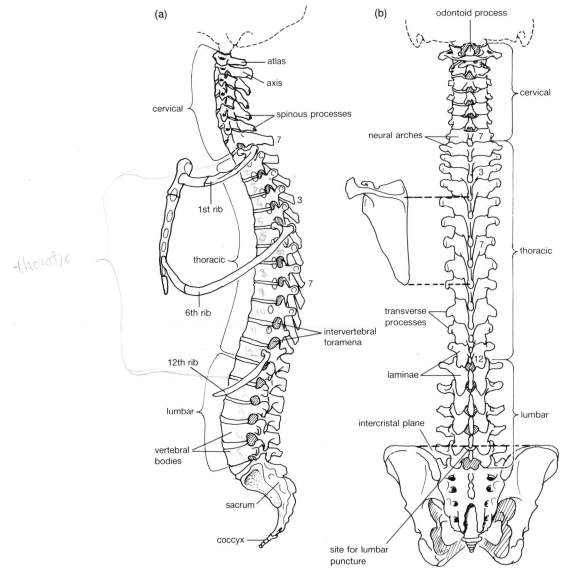

thoratic

8.2
Vertebral column: (a) lateral view,
(b) posterior view.

with his thoracic spine pressed against a wall, holding the head in the **natural erect posture**. The degree of kyphosis can be assessed by the horizontal distance from the wall to the tragus of the ear (the cartilaginous lump just in front of the external auditory meatus).

What is the wall-to-tragus distance in
a) your partner   cm?
b) yourself   cm?

Alterations of the normal curvature may occur (**8.3**, **8.4**). Any deviation to the right or left of the midline vertical constitutes a deformity known as a scoliosis (**8.3**, **8.4a**). Minor degrees of 'postural' scoliosis are unimportant and may disappear on flexion of the spine. 'Structural' scoliosis is a permanent deformity. Scoliosis may result from

maldevelopment (a hemivertebra—failure of one side of a vertebra to develop or presence of a supernumary half-vertebra on one side), or from contracture of one side of the chest wall due to underlying lung disease. Most scolioses, however, are of unknown origin and involve rotational displacement of all the adjacent vertebrae, the bodies of the vertebrae pointing to the concavity of the curvature. The normal kyphosis may be increased (**8.4b**), particularly if the bones become weak through lack of calcium, as may occur in the elderly. A severe scoliosis may be associated with an increased kyphosis which severely diminishes the volume of the thoracic cavity and causes lung and heart problems. A localized injury or collapse of a single vertebral body can produce an angular kyphosis, the visible bony apex of which is called the kyphus.

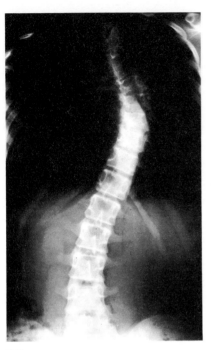

8.3
Radiograph (AP) showing scoliosis.

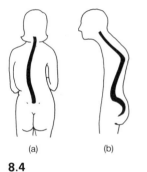

(a)   (b)

8.4
(a) Scoliosis; (b) Kyphosis.

Identification of vertebral levels on the back

Run your finger down the midline of your partner's back, starting at the base of the skull. Upper cervical spinous processes lie deeply within the neck muscles; the first easily felt spinous process (the '**vertebra prominens**') is usually C7 or T1. Draw a line joining the uppermost parts of the iliac crests (intercristal plane) and note that it crosses the midline between the spinous processes of L4 and L5. The scapula is very mobile but, with the arms at rest by the side, the spine of T3 is usually at the level of the base of the spine of the scapula, and the spine of T7 is usually at the level of the lower angle of the scapula.

Qu. 1A *Locate the spinous process of L1. The spinal cord ends at this level in the adult although it extends to L5 in the newborn child. Why is this?*

Feel the prominence of muscular columns (erector spinae) which lie on each side of the lumbar vertebral spines.

Qu. 1B *When your partner stands on one leg, these muscular columns become particularly prominent on one side. On which side, and what is the explanation?*

Note also the dimples that overlie the posterior superior iliac spines of the pelvis.

Movements of the spine

Next, assess the **range of movement** of the spine:

Flexion
Ask your partner to stand, knees straight, and then bend forward to try to touch the ground with the tips of his fingers. The distance between fingertip and ground provides a rough measure of the flexion achieved. As a true measurement it is complicated by the degree of hip flexion and the tension of the hamstring muscles.
Ask your partner to stand erect and measure the distance between the spine of S1 and the vertebra prominens. Repeat this with your partner flexed i.e. touching his toes. Comment on the results.

S1 to vertebra prominens in erect posture   cm
in full flexion   cm

Lateral flexion
This is assessed by recording the level attained by the tips of the fingers reaching down the side of the thigh to the knee, without rotation of the trunk or flexion of the knees.

Extension and rotation
These are less easy to measure and are usually assessed by eye.
From examination of your partner assess which regions of the spine contribute most to
a) flexion and extension
b) rotation
c) lateral flexion

Development of the spine (8.5)

The first axial element to be formed is the **notochord** which underlies the neural plate/tube (future spinal cord). This mesodermal structure characterizes us as members of the chordate phylum. The notochord contributes mainly to nucleus pulposus of the intervertebral discs. On each side of the axis formed by the neural tube and notochord the mesodermal **somites** form in rostrocaudal sequence (p. 142). The ventromedial part of each somite loses its epithelial arrangement to form a **sclerotome**, which will give rise to cartilaginous and bony elements. Cells of the

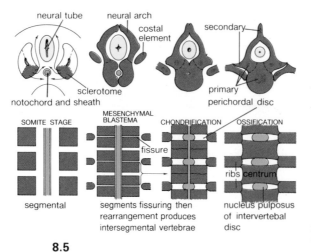

8.5
Stages of development of the spine.

sclerotome migrate to surround the notochord and neural tube to form a mesenchymal precursor of the vertebral column. Other somite cells form the **myotome** which gives rise to the segmental muscles of the spine.

The mesenchyme of the sclerotomes is segmentally arranged; it is subdivided by intersegmental (intersomitic) fissures which correspond to the original boundary between somites, and intrasegmental fissures which separate the rostral and caudal halves of each sclerotome. Each intrasegmental fissure lies opposite the middle of the overlying myotome and marks the position at which the intervertebral disc will develop. Above and below the centre of each somite a **perichordal disc** forms, derived from a local thickening of the notochord and a condensation of sclerotome-derived cells. The perichordal disc is thought to contribute to both the intervertebral disc and the centra. It is generally assumed that the centra of vertebrae form from the adjoining halves of two adjacent sclerotomes. However, there is no direct experimental evidence for this, the conclusion being based almost entirely on histological observation. An alternative is that individual sclerotomes correspond to individual vertebrae and that, when the sclerotome cells migrate toward the notochord, they also move rostrally by about one half-segment.

Whatever the mechanism, the development of the perichordal discs out of phase with the original somites allows the correct alignment of several important features. The segmental spinal nerves pass out between individual vertebrae; the intersegmental arteries (e.g. intercostal arteries, lumbar arteries) lie on the vertebral centra, and the axial muscles derived from each myotome span two adjacent vertebrae, allowing movement of one vertebra on the next.

Chondrification of the centra starts at about the 6th week in utero, from two centres which rapidly fuse. Each half of the neural arch chondrifies from a centre at its base. Costal elements chondrify separately. The neural arch chondrification does not unite in the midline until the 4th month in utero.

Ossification of typical vertebrae from **primary centres** starts at 7–8 weeks in utero. The ossification centres are in the same position as the chondrification centres. The neural arches unite as bone during the first year, but the neurocentral joint between the arch and the centrum persists for the first few years of life. At puberty **secondary centres** appear in the spinous process and transverse process, and two circumferential, annular epiphysial discs form at the cranial and caudal ends of the vertebral bodies.

Particular ossification patterns occur in the atlas, axis, and lumbar vertebrae. The anterior arch of the atlas is formed from the **hypochordal bow**—tissue which unites the costal elements anterior to the centra. The centrum of the 1st cervical somite forms the odontoid peg of the axis. The pattern in the sacrum is essentially similar to that in typical vertebrae, except that the vertebrae fuse with each other and with the costal elements which form the alae.

Congenital anomalies of the spine

Congenital anomalies of the spine are relatively common. The spine is a functional unit made up of many vertebrae each of which is formed from a number of elements which must segment, chondrify, and ossify correctly. A defect in a single vertebra such as a hemivertebra (arrows, **8.6**) can so disturb the normal anatomy as to make the spine **unstable** so that when the stresses of postnatal life are imposed, progressive deformation will result. Other anomalies such as fusion of two or more

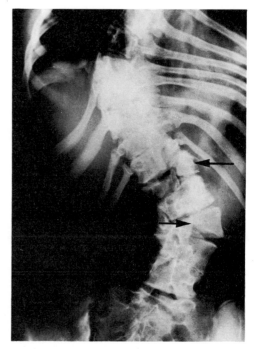

8.6
Congenitally maldeveloped spine showing hemivertebrae.

adjacent vertebral bodies may produce a minor loss of movement which can easily be compensated for, but they are essentially **stable** and do not lead to progressive deformity. Many spinal anomalies are associated either primarily or secondarily with nervous system defects and damage. **8.7a** is a radiograph of a young patient with spina bifida. The neural arches of L2 and L3 (dotted) can be seen, but those of L4, L5, and S1 are missing (arrows). **8.7b** is a horizontal MRI through such a defect, showing the gap in the neural arch through which the meningeal sac has herniated (arrow). When considering specific deformities think of their effects not only on the mechanics of the spine but on the thoracic cavity and respiration, and on the spinal cord and nerves.

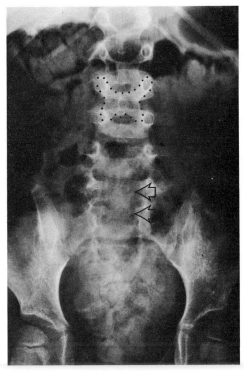

(a)

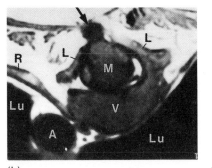

(b)

8.7
(a) Radiograph spina bifida; (b) Horizontal MRI of spina bifida showing protruding meningeal sac (arrowed). V body of thoracic vertebra; L laminae of vertebra; M meningeal sac containing spinal cord; R rib; A aorta; Lu lungs.

The bony skeleton (8.2, 8.8–8.13)

Examine vertebrae from different regions of the spine.

On each vertebra you should identify:

● the **centrum** or body of the vertebra

● the **vertebral arch**. Each arch comprises, on each side, a **pedicle** attached to the vertebral body and a broad **lamina**. Together with the vertebral body, the arch encloses the **vertebral foramen**.

● a **spinous process** which projects from the arch dorsally in the midline

● two **transverse processes**

● paired **superior** and **inferior articular facets**, which project from the pedicles so that the vertebral arch articulates with the vertebra above and below.

Note the shape and orientation of the superior and inferior articular facets in the different regions. These form synovial joints at which a small amount of gliding movement occurs.

Qu. 1C *How does the shape of the articular surfaces relate to the movement possible in a particular region of the spine?*

Multiple small vascular foramina can be found on the bone. The largest are seen on the dorsal surface of the centrum where basivertebral veins drain from the red marrow to the internal vertebral venous plexus (p. 150).

Costal elements develop in all regions but normally form separate ribs only in the thoracic region. In the cervical region they form the part of the transverse process anterior to the foramen transversarium; in the lumbar region they are incorporated into the transverse process. Abnormally separate ribs may develop in either region but are most common at C7 and L1.

Qu. 1D *Would you expect a greater proportion of people with a C7 accessory rib or an L1 accessory rib to complain of symptoms, and why?*

Cervical vertebrae:

C1—atlas (8.8): The atlas does not have a centrum; during development this became the odontoid process of the axis. Its short **anterior arch** has an anterior tubercle to which the anterior longitudinal spinal ligament is attached at its apex, and an articular facet for the dens. The larger **posterior arch** is grooved on its upper surface by the vertebral artery and a posterior tubercle represents the spinous process. The **lateral masses** are each pierced by a foramen transversarium for the vertebral artery. The attachment for the transverse ligament of the atlas can be seen in the vertebral canal and the **transverse processes** are prominent. Large superior facets articulate with the occipital condyles.

C2—axis (8.9): The axis is easily distinguished by the **odontoid peg (dens)** which bears articular

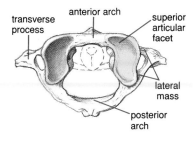

8.8
Atlas (C1), from above.

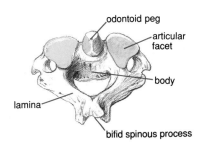

8.9
Axis (C2), from above.

facets for the anterior arch of the atlas on its anterior surface, for the transverse ligament of the atlas on its posterior surface, and for the attachment of the alar ligaments above. The **body** has prominent oval articular facets above which rotation on the atlas occurs. The **laminae** are very thick and the **spinous process** large and bifid as in most other cervical vertebrae.

Cervical vertebrae C3–6 are the 'typical' cervical vertebrae. Each has a relatively small **body**, a **transverse process** pierced by a foramen transversarium with an anterior tubercle (costal element) and posterior tubercle. Note the lateral lips on the upper surfaces of the bodies and the shape and orientation of the articular facets. The **spinous processes** are bifid for the attachment of the ligamentum nuchae.

Cervical 7 vertebra (8.10) is the 'vertebra prominens'—the first that can easily be felt as the examining hand descends the neck. Its spinous process is horizontal and not bifid, and the vertebral artery does not traverse its (often narrow) foramen transversarium.

Thoracic vertebrae (8.11) have bodies of increasing size which typically bear upper and lower hemifacets for the heads of the ribs. The vertebral foramen is relatively small; the laminae are thick, broad, and overlapping; the spinous processes are long and slant downward and backward; the transverse processes are substantial and most bear facets for articulation with the tubercles of their numerically corresponding ribs.

Note also the following distinguishing features:
T1 has a complete upper facet on the body and a
   thick, horizontal spine
T9 often does not articulate with the 10th rib
T10 articulates with only the 10th rib
T11 articulates with only the 11th rib and lacks
   facets on its transverse processes
T12 articulates with only the 12th rib and lacks
   facets on the transverse processes.
Interlocking articular facets of the lumbar type start at the lower border of T11 or T12.

Lumbar vertebrae (8.12) have large, deep bodies; short, stout pedicles; a quadrangular spinous process and long thin transverse processes (costal elements); the articular processes are of the interlocking type.

● L5 has a short, thick, conical transverse process that gives attachment to the ilio-lumbar ligament which connects it to the pelvis.

The sacrum (8.13) is large and triangular, and normally consists of 5 fused vertebrae. The base articulates with L5 at the sacrovertebral angle. Anteriorly it projects as the sacral promontory. The pelvic surface is pierced by 4 anterior sacral foramina, lateral to which the fused costal elements form the lateral part of the sacrum with a smooth anterior ala (= wing). Dorsally the median sacral crest represents the (fused) spinous processes and 4 dorsal foramina can be seen. Laterally the fused

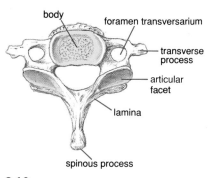

8.10
Cervical vertebra C7, from above.

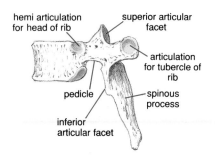

8.11
Typical thoracic vertebra (T5) from side.

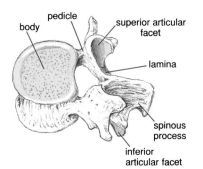

8.12
Typical lumbar vertebra (L4), from above and side.

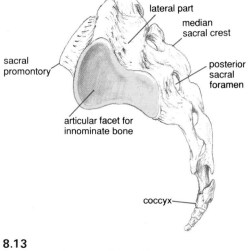

8.13
Sacrum and coccyx from side.

transverse and costal elements form the L-shaped articular surface in its upper part. The whole sacrum forms a gently curving roof and posterior wall of the pelvis. Below it the coccyx consists of 3–5 rudimentary vertebrae.

The **number of vertebrae** in different regions may vary, usually by only one element. Most commonly this occurs in the lumbosacral region, owing to the complete or incomplete sacralization of a lumbar vertebra.

Examine in particular the articulation between L5 and the sacrum (see **8.14**). The upper surface of the body of S1 slopes downward and forward when the trunk is erect. Forward movement of L5 on S1 is normally prevented by the articular facets of the sacrum which lie anterior to those of L5 and by the integrity of the intervertebral disc between the vertebrae.

There is a condition known as spondylolisthesis (vertebral slipping), which may begin when an infant starts to toddle, in which a defect such as a fatigue fracture can occur between the superior and inferior facets of L5, a region known as the pars interarticularis. The body of L5, carrying the spinal column above it, can then move forward, leaving the posterior facets, laminae, and spinous process behind. As displacement of the spine continues, compression of the nerves in the sacral canal (particularly S1) can occur.

B. Radiographs

Now examine the radiographs of the different regions of the spine (**8.14–8.21**) and identify all the components you have located on the dry bones. An oblique view of the lumbar spine (**8.22**) presents a 'Scottie dog' appearance in which the dog's nose is the transverse process, its ear the superior articular facet, and its eye the pedicle. The collar is the 'pars interarticularis'—the anterior part of the lamina.

C. Prosections

The ligaments of the spine (8.23, 8.24)

The individual vertebrae are held together by a series of **long** and **short** ligaments. On prosections identify and examine:

Short:
interspinous ligaments between adjacent spinous processes,
intertransverse ligaments between adjacent transverse processes,
ligamenta flava between the laminae of the neural arches; these are yellow, elastic ligaments.

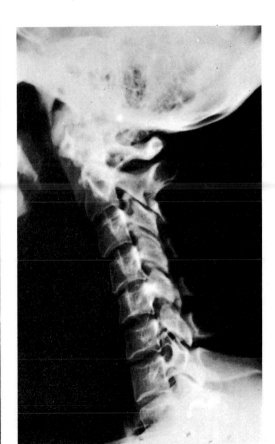

8.14
Cervical spine (AP).

8.15
Cervical spine (lateral).

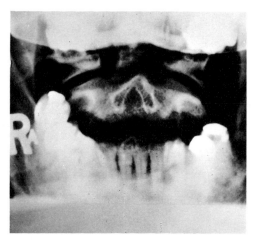

8.16
Axis viewed through open mouth.

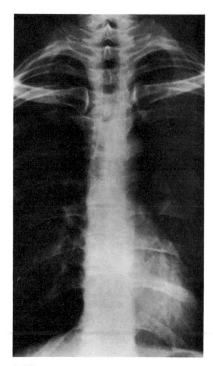

8.17
Thoracic spine (AP).

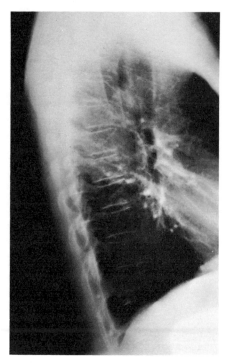

8.18
Thoracic spine (lateral).

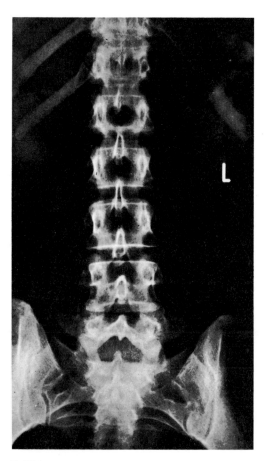

8.19
Lumbar spine (AP).

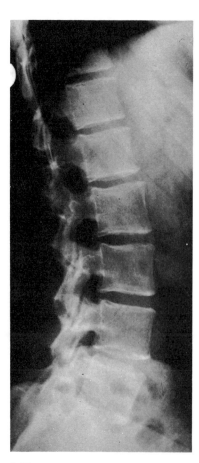

8.20
Lumbar spine (lateral).

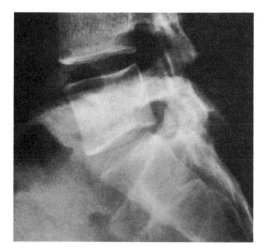

8.21
5th lumbar vertebra.

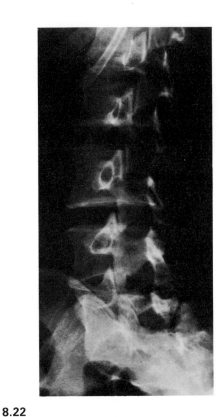

8.22
Lumbar spine (oblique; to show 'Scottie dog').

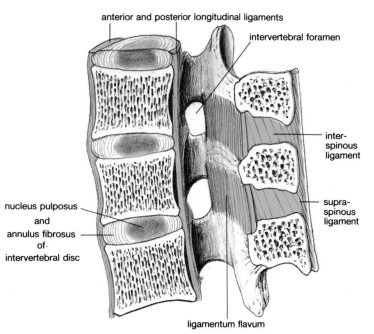

8.23
Sagittal section through lumbar spine.

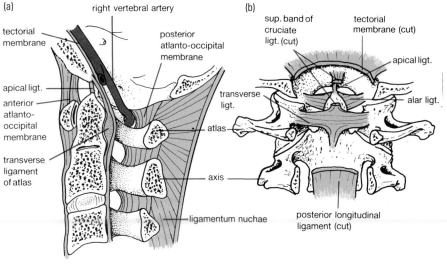

8.24
(a) Sagittal section through cervical spine. (b)
Ligaments of occipito-atlanto-axial region
(dorsal view).

Long:

anterior longitudinal ligament on the anterior aspect of the centra,

posterior longitudinal ligament on the posterior aspect of the centra,

supraspinous ligament over the tips of the spinous processes. This becomes the **nuchal ligament** between C7 and the skull.

In the region immediately beneath the skull, both vertebrae and ligaments are specialized in relation to the movements that occur between the occiput and the atlas, and between the axis and the atlas. Move your own head and study the bones to discover the movements of this region. Nodding flexion/extension occurs at the atlanto-occipital joint with very minor lateral flexion possible. The atlanto-axial joint is specialized to allow rotation; other movements being prevented by the restriction of the odontoid peg within the anterior arch of the atlas by the transverse ligament.

On a prosection (**8.24**) identify the:

anterior atlanto-occipital membrane—continuous with the anterior longitudinal ligament.

tectorial membrane—continuous with the posterior longitudinal ligament.

posterior atlanto-occipital membrane—the homologue of ligamenta flava.

cruciate ligament—this comprises the transverse ligament of the atlas with extensions passing upward and downward.

alar ligaments which extend from the top and sides of the odontoid process laterally to the margin of the foramen magnum.

apical ligament—this thin band, formed from the notochord, may have a small bone, a pro-atlas, within it.

Qu. 1E *What is the function of the transverse ligament of the atlas?*

Qu. 1F *What movements do the alar ligaments restrain?*

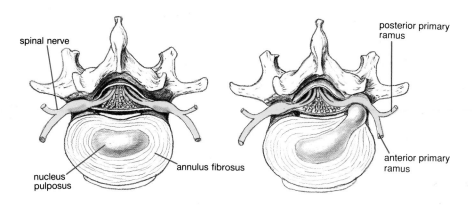

8.25
(a) Lumbar intervertebral disc viewed from below. (b) Protrusion of nucleus pulposus ('slipped disc').

Intervertebral discs (8.25)

Examine prosected intervertebral discs. Identify the fibrous outer ring, the **annulus fibrosus**. Although it may be difficult to see them, the annulus is composed of a series of layers and adjacent layers of the fibres pass obliquely in different directions, giving great torsional strength. Identify the **nucleus pulposus** lying within the annulus. In the child the annulus is thin and the nucleus central. With ageing the annulus becomes thicker at the front and the nucleus comes to lie more posteriorly in the disc. For this reason any herniation (protrusion) of the nucleus pulposus—

known as a 'slipped disc'—is likely to occur in a posterior or posterolateral direction.

Intrinsic muscles of the spine

There are essentially three groups of intrinsic spinal muscles:

a) **flexors (8.26)**—this relatively weak anterior group is found in the neck (longus colli) and extends from T4 to the base of the skull (longus capitis) with attachments at all levels. In the lumbar region psoas major acts as a more powerful flexor. These flexors are supplied segmentally by anterior primary rami of spinal nerves. Muscles placed anteriorly in the body but separated from the spine also act as powerful spinal flexors; examples of such muscles are sternomastoid in the neck and rectus abdominis in the trunk.

b) A **lateral** group—the **scalene** muscles. These run from cervical transverse processes to ribs, produce lateral flexion of the neck, and act as 'guy lines' supporting the head on the trunk (see Vol. 3).

c) **Postural muscles** and **extensors (8.27)**— this large group of muscles lies posteriorly and is arranged as a series of long, medium, and short muscles. The shortest muscles, i.e. those between individual transverse processes and between individual spinous processes, lie most deeply and are mainly **postural** in function, producing only a little extension and rotation. Medium-length muscles lie superficial to these and run from transverse processes to spinous processes of vertebrae adjacent or of more distant regions. The longest and most superficial muscles form the **erector spinae**. This group is firmly anchored to the dorsum of the sacrum and is the prime mover in extension of the spine. Fibre bundles are attached to spinous processes, transverse processes, and costal elements. Some of its components extend to the occiput and thus produce extension of the head. In the region immediately beneath the skull, a specialized group of **suboccipital muscles (8.28)** controls extension and rotation at the atlanto-occipital and atlanto–axial joints respectively (see Vol. 3).

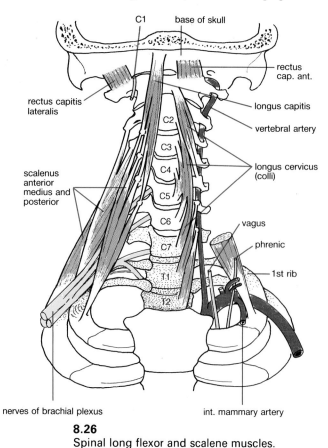

8.26
Spinal long flexor and scalene muscles.

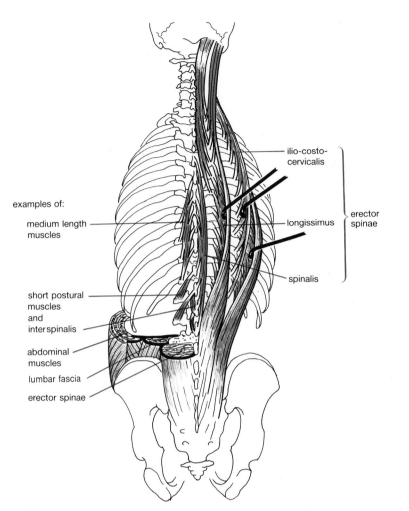

8.27
Extensor muscles of the spine.

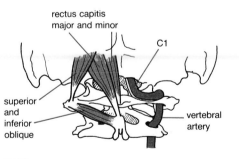

8.28
Suboccipital muscles.

All the muscles of the extensor group are supplied by the **posterior primary rami** of the spinal nerves which also supply the skin of the back.

Contents of the spinal canal
(8.29)

Within the **spinal canal** lie the spinal cord and spinal nerve roots, surrounded by protective membranes, the **meninges**. The outer tough fibrous layer is called the **dura mater** and this extends from the head throughout the length of the spine to S2 where it narrows to a thin cord. In the **extradural space** between the bone and dura, lies an **internal vertebral venous plexus** supported by a little fat. This plexus drains the marrow in the vertebral bodies via basi-vertebral veins which emerge from the dorsal surfaces of the centra, and it communicates with a plexus outside the spinal column, the **external vertebral venous plexus**, via the intervertebral foramina. Blood flow in this system offers a venous pathway to the heart which can bypass the abdominal cavity, particularly when intra-abdominal pressure is raised. It is also an important route for the spread of (pelvic) malignancy since it is without valves and is connected to veins of the pelvis, the flow of blood eddying to and fro through the external and internal plexuses with, for example, changes of posture.

The dura mater is lined by a cobwebby membrane, the **arachnoid mater**, within and by which the spinal cord is suspended in **cerebrospinal fluid** (CSF). The CSF-filled **subarachnoid space** ends at S2 level.

Samples of CSF can be obtained by inserting a needle into the subarachnoid space in the lower

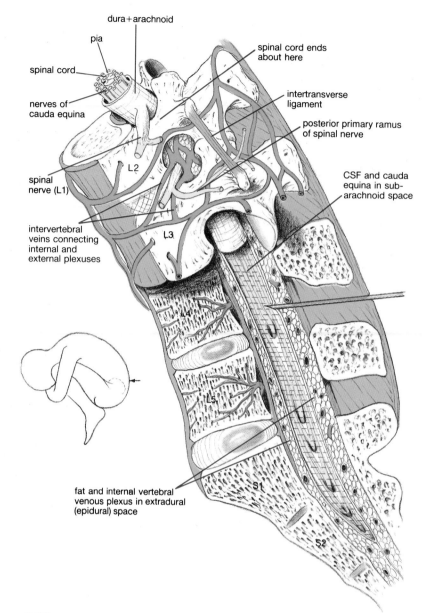

8.29
Spinal canal and its contents.

lumbar region ('lumbar puncture'). Withdrawing CSF from a patient who has raised intracranial CSF pressure can have fatal consequences. Therefore, as a preliminary, the pressure of the CSF is assessed by examining the optic disc of the eye (see Vol. 3). The skin over L4 and L5 is anaesthetized and the patient lies on his side, curled up as much as possible. A lumbar puncture needle is introduced in the midline, pointing forward and slightly toward the head, between the spines of L4 and L5 vertebrae (the intercristal plane), through the interspinous ligament, through the ligamentum flavum, and so into the spinal canal. Advancing further, the needle will pierce the dura and arachnoid membranes and enter the CSF-containing subarachnoid space which, below the level of L1, contains the mass of lower lumbar and spinal nerve roots known as the

cauda equina. Before any fluid is withdrawn for analysis the pressure is measured with a manometer. Take an articulated lumbar spine and pass a needle as you would in a lumbar puncture, mentally noting the structures which are traversed.

Qu. 1G *Why is a patient undergoing lumbar puncture asked to curl up as much as possible?*

This access to the spinal dural sac also permits:

• an anaesthetist to inject local anaesthetic into the CSF around the roots of the cauda equina, to allow surgery to be performed on the pelvis (e.g. during childbirth) or legs. Such **spinal anaesthesia** is especially useful where general anaesthesia is contraindicated. Extradural application of anaesthetic around emerging spinal nerves (**epidural anaesthesia**) can also be used.

● a radiologist to inject non-irritant radio-opaque dye into the cerebrospinal fluid. Radiologic examination (a myelogram) (**8.30**) then reveals the shape of the dural sac, and the outline of the dural 'sleeves' at the origin of the roots as they pass through the dura. The patient is tilted to allow the viscous dye to reach the appropriate level. A distortion of the normal outline of the dural sac can reveal, for instance, protrusion of an intervertebral disc (arrow). **8.1.31** is a sagittal MRI of the lumbosacral region in which the L5/S1 disc is protruding (arrow) and compressing the anterior aspect of the meningeal sac and the nerve roots within it.

The spinal cord and spinal nerve roots are usually well protected by the vertebral bodies and by the joints and ligaments which connect them. However, in severe injuries (a car accident, fall from a horse, roof fall in a mine, etc.) the spine may break, especially at the junctions of relatively mobile and immobile segments (the lower cervical and thoracolumbar regions). One vertebra and the spinal column above it can shear away completely from the adjacent vertebra and spinal column below. It is important for the doctor attending such a patient to appreciate the damage that can be caused by movement after, as well as at the time of, the accident.

Qu. 1H *If the injury was between C6 and C7, such that the spinal cord was transected, what would be the results? Would the patient be able to breathe? What arm movements would he still be able to perform?*

If the injury were of a similar nature but between T12 and L1, there could be damage to either or both of the spinal cord (its terminal sacral part) and the nerve roots that are passing down to their respective intervertebral foramina. It is important to distinguish between nerve root and spinal cord damage since, in the former, there is the possibility of some recovery, as fibres can regrow down Schwann cell sheaths. With cord injuries the damage will be permanent.

Qu. 1I *How will the results of an injury at T12/L1 differ from those of an injury at C6/7?*

If the level of transection is below L1, then only nerve roots will be damaged.

Because the spinal canal is surrounded by bone it follows that any structure within the canal can expand only at the expense of the other contents. An expanding tumour or disc prolapse in the canal can very quickly cause compression of the spinal cord or its roots. One of the most common causes of nerve damage in the spinal canal is the posterior protrusion of part of the nucleus pulposus through a degenerate annulus fibrosus (**8.25**). The discs most commonly involved are those between L4/L5 and L5/S1. The protrusions usually occur posteriorly, but to one or other side of the posterior longitudinal ligament. For effects in the lower limbs see p. 137. Painless retention of urine also occurs from interruption of bladder innervation.

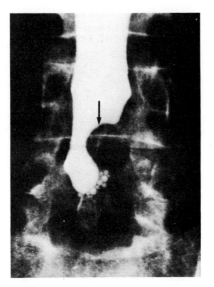

8.30
Myelogram showing protruding lumbar intervertebral disc (arrowed).

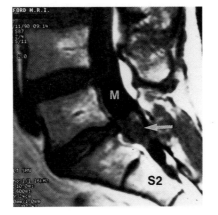

8.31
MRI of lumbar spine (sagittal section) showing protrusion (arrow) of L5/S1 disc. Note the ending of the CSF-filled meningeal sac (M) opposite the lower border of S2.

Examine the skeleton and note the relationships of the intervertebral foramina and the anterior and posterior sacral foramina. **8.25** shows a transverse view of the lumbar spinal canal. A disc protrusion between L4 and L5 impinging on the L5 nerve root is shown; the lower lumbar region is the most common site of disc protrusion. Very large disc protrusions may affect several nerve roots together.

Qu. 1J *What would be the effect of such a compression of an L5 nerve root?*

Disc lesions may occur in the cervical region (and very rarely in the thoracic region). In these regions not only nerve roots but also the anterior aspect of the spinal cord may be compressed. The anterior part of the cord contains motor nerve cell bodies and descending tracts which control them.

8.32
Straight leg raise test.

Spinal nerves, formed by the union of anterior and posterior nerve roots, pass out of the spinal canal through the intervertebral foramina.

Examine the boundaries of an intervertebral foramen: they are formed by edges of the centrum and pedicle, the intervertebral disc, and the facets of the posterior intervertebral joints. Expansion of any of these caused, for example, by postero-lateral disc protrusion, arthritic joint changes, or bony overgrowth, will encroach on the intervertebral foramen and damage the spinal nerve within.

Lay your partner supine on a couch and raise one of his legs, making sure that the knee is fully extended. There is a limit to which this procedure (the straight leg raise) can be performed without causing discomfort. As the straight leg is raised the spinal nerve roots which contribute to the sciatic nerve are stretched and move several millimetres through their respective intervertebral foramina, thereby pulling the dural sac downward and to one side. If the nerve root is already being compressed this test will increase the pain. Dorsiflexion of the foot during a straight leg raise will have the same effect for similar reasons. The extent to which a 'straight leg raise' can be achieved in comfort can be used to assess progress of a disc lesion (**8.32**). A similar test of the femoral nerve can be done by getting the patient to lie in a prone position and flex the knee.

Sacroiliac joints

The wedge-shaped sacrum articulates with the innominate bones of both sides to form the bony pelvis which supports and protects the pelvic viscera and, in the standing position, transmits the weight of the trunk via the acetabulum to both lower limbs. The **sacro-iliac joint** which unites them is, perhaps surprisingly, a synovial joint, but one at which essentially no movement normally

occurs. The articular facets on the sacrum and ilium are ear-shaped (auricular). The bony facets and the hyaline cartilage that covers them are reciprocally irregular so that they interlock when held close together, and some fibrocartilage extends between the articular surfaces. The articular surfaces are held together by **ventral** and **dorsal sacro-iliac ligaments** that extend upward and laterally from the alae of the sacrum to the iliac bones, and by the very strong **interosseous sacro-iliac ligament** that fills the irregular space above and behind the joint cavity. These are reinforced by the **iliolumbar ligament** which extends from the short, conical transverse process of the fifth lumbar vertebra to the adjacent part of the iliac crest, and by the **sacrotuberous** and **sacrospinous ligaments** (p. 107). The interlocked articular surfaces and strong ligaments normally prevent the upper part of the sacrum from being forced downward and forward by the weight of the trunk. The sacrum can dislocate forward within the pelvis only if force is applied after the ligaments have become lax. Pelvic ligaments often become lax at around the time of birth. Dislocation of the sacroiliac joint is therefore sometimes seen in underdeveloped countries in women who return to hard, stooping work in the fields very soon after giving birth.

Requirements:

Articulated skeleton; separate vertebrae of the different regions; separate ribs

Prosections of the spine; ligaments; intervertebral discs; intrinsic muscles

Prosections of the joints, ligaments, and muscles of the occipito–atlanto–axial region

Prosections of the dorsal aspect of the spine with laminae removed to expose the meningeal coverings of the spinal cord and cauda equina

Radiographs of different regions of the spine.

CHAPTER 9

Human upright stance, sitting, and locomotion

Man's upright stance has evolved by the balancing of the trunk on the lower limbs, and the head on the trunk. The lower limbs became specialized to support the entire weight of the body and to provide for most forms of locomotion, thus freeing the upper limbs for gesticulation and manipulating a wide range of objects.

Standing rest position

In the upright 'standing rest' position the weight of the body is evenly distributed between the two legs and the arms hang loosely on either side although, in fact, most people shift their body weight alternately from one leg to the other. The weight of the body above the pelvis is transmitted through the bones and intervertebral discs of the flexible vertebral column to the sacrum which is therefore thrust downward and forward between the two sides of the pelvis. The downward force is countered by the interdigitating surfaces of the sacroiliac joints and the strong sacroiliac ligaments while the forward rotatory force is countered by the sacrotuberous and sacrospinous ligaments (**9.1a**). Weight transmitted from the sacrum to the pelvis is transmitted via the femur and tibia to the talus at the arches of the feet, and thence obliquely forward to the heads of the metatarsals and backward to the posterior tubercles of the calcaneum.

The body's **centre of gravity** in the standing rest position lies immediately in front of the 2nd sacral segment, but it cannot be static because the body sways constantly, though to a small extent and primarily in the sagittal plane, its lowest (transverse) axis being the ankle joints. The vertical **central line of gravity (9.1b)** passes downward in a plane which passes through or just behind the head of the femur, and through the patella (i.e. just in front of the knee) to intersect the arch of the foot well in front (2–5 cm) of the ankle joint, the femur and tibia being angled forward at 5° from the vertical. The weight of the body therefore exerts a forward-rotatory force at the ankle. Electro-myographic (EMG) studies of muscle activity in the leg show that, to counter this force, soleus is active when standing quietly and is therefore an important postural muscle.

At the knee, where the line of gravity passes close to the axis of the joint, less muscular counterbalance is required than at the ankle. The knee is not locked in full extension in the standing rest position but is near its close-packed position and you can demonstrate that quadriceps is inactive because the patella can easily be displaced laterally. The hamstrings are also inactive and thus no obvious muscle contraction stabilizes the knee. The small forward-falling force at the knee joint is largely taken by the posterior and collateral ligaments, and by the large contact area between the condyles and menisci. If, however, the knee is either fully extended or slightly flexed then both quadriceps and the hamstrings become active.

The line of gravity passes just behind (or even through) the axis of the hip joint and so there is a small backward-falling (extensor) force on the joint. As at the knee, this requires little muscular counterbalance (by psoas and iliacus) and is largely taken by the strong iliofemoral ligament. EMGs indicate that gluteus medius is also active, its anterior fibres contracting to prevent extension and its posterior fibres contracting to prevent flexion as the trunk sways slightly about the transverse axis of the hips. The deep, short muscles around the hip (e.g. obturator externus) may also play a role in hip stability but EMG evidence for this is difficult to obtain.

The centre of gravity of the trunk in an upright standing rest position is just in front of the 11th thoracic vertebra, and thus tends to flex the spine. This is countered by activity in the small deep postural components of the spinal extensor muscle groups. Even a small forward inclination is sufficient to require increased activity in erector spinae.

The line of gravity of the head passes in front of the atlanto-occipital joints and the flexion force on this joint is well demonstrated by a drooping head when one 'nods off' during a lecture! The rapid contraction of skull extensors as they reflexly return the head to the horizontal can be equally embarrassing!

Sitting

There are a variety of different sitting positions, which depend in part on the extent of external support provided (bench, chair with back, etc.). If the trunk is upright, or sloping backward then the centre of gravity is above the ischial tuberosities which provide the principal bony contact with the surface. If however the trunk is bent forward the centre of gravity lies well anterior to the tuberosities somewhere above the middle of the thighs and there is then a considerable flexor moment on the hips and spine. The extent to which the pelvis is flexed or extended on the thighs also varies. Because of the considerable problems humans have with their

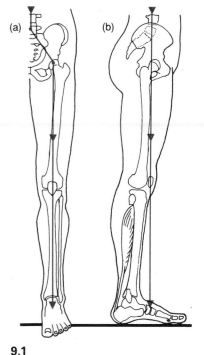

9.1
Lines of gravity: (a) anteroposterior view; (b) lateral view (most of the fibula has been removed).

lumbar intervertebral discs, and the fact that many spend much of their life sitting, there has been special interest in the relationship between disc loading and the sitting posture. Although the deep back muscles are more active in the upright positions, the load on the dorsal aspect of the L4–5 intervertebral disc (where rupture usually occurs; p. 151) is greater in forward-hunched sitting positions. This is a complex subject and the load on the back also depends on the position of the arms and whether or not they are supported. There is probably no one ideal sitting position, but a number from which to choose. Also small shifts of position alter the loads and tensions and relieve the strain.

Locomotion

Man achieves locomotion in many different ways, depending on the circumstances (e.g. crawling, hopping, sliding); but the two most common bipedal forms of locomotion are both variants of the **striding gait**—walking and running. Other animals may walk on two legs intermittently, but the striding gait is unique to Man. The basis of the striding gait is that, while one leg supports and propels the body forward (**support phase**), the other swings forward free of the ground (**swing phase**) to a new position where it, in turn, supports and propels the body as the cycle is repeated (**9.2**).

Walking is an exceptionally energy-efficient form of locomotion, in part because the centre of gravity of the body is disturbed very little (**9.3**, **9.4**), and in part because of the storage of elastic recoil energy in the muscles and tendons. The more rapid but less energy-efficient running also evolved to carry our ancestors away from danger or towards prey. The speed of locomotion can, of course, be varied in both walking and running.

If you walk slowly or watch someone walking, you will see that the support phase of one leg does not end immediately the opposite foot strikes the ground (**heel strike**). Rather, for a brief period, both feet are simultaneously in contact with the ground (**double-support phase**) (**9.2b**). As

walking quickens, so the double-support phase shortens until running is achieved, in which, for a period, neither foot is on the ground (**float phase**).

Stand with your feet together at rest, and then slowly start to walk forward. Walking is initiated by relaxation of soleus which causes the ankle joints to dorsiflex passively because the line of gravity runs in front of the ankle (**9.2a**). The weight is transferred from the heel to the toes, especially the great toe, along the lateral side of the foot. The limb which will enter the swing phase is then actively flexed at the hip to propel the limb forward and flexed at the knee and dorsiflexed at the ankle to take the foot clear of the ground. As the body continues to swing forward on the supporting leg, the knee of the swinging leg is increasingly extended and hip flexion is braked as heel strike approaches. At the same time the hip and knee of the supporting limb become fully extended (**9.2a, d**) so that the limb becomes, momentarily, a rigid prop as ankle and then first metatarso-phalangeal joint are forcibly (plantar-) flexed to provide the propulsive thrust (**9.2b**).

As each leg is swung forward, the *pelvis* also rotates about a vertical axis to increase the length of the stride (**9.3**). Therefore, in order that the feet shall keep pointing forward during walking, each lower limb is laterally rotated on the pelvis during the swing phase, and medially rotated during the stance phase. When a leg is swung forward free of the ground it can no longer support the body weight, which is taken by the support-phase leg. To prevent the pelvis from dropping on the unsupported side, the pelvis is abducted on the femur of the supporting leg. This also helps to enable the swinging leg to move forward clear of the ground (**9.4**; see p. 108).

It is not only the lower limbs and pelvis which move during walking. The trunk (carrying the body's centre of gravity with it) moves to the side of the supporting leg by lateral flexion of the spine. Also, the shoulder girdle swings clockwise around the vertical axis so that, as the pelvis rotates in one direction, the shoulders swing in the opposite way to balance the trunk (**9.3**). The arms are also swung to act as balancers and to increase the impetus of movement, especially in running and jumping.

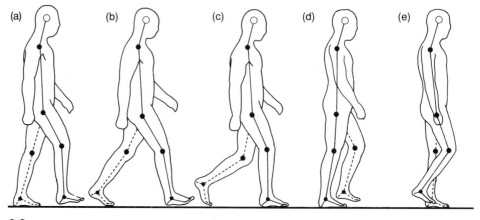

9.2
Walking cycle.

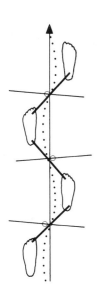

9.3
Movement of the transverse axis of the pelvic girdle (thick line) and shoulder girdle (thin line), and centre of gravity (dotted line) in relation to foot positions when walking in the direction of the arrow.

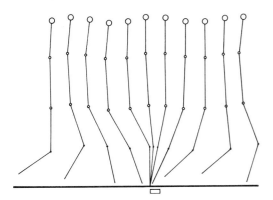

9.4
Vertical movements of the hip, shoulder and head during a walking cycle. (Support phase marked by a block.)

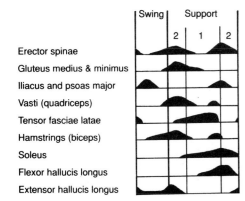

	Swing	Support		
		2	1	2
Erector spinae				
Gluteus medius & minimus				
Iliacus and psoas major				
Vasti (quadriceps)				
Tensor fasciae latae				
Hamstrings (biceps)				
Soleus				
Flexor hallucis longus				
Extensor hallucis longus				

9.5
Activity of various muscles and muscle groups during the swing and single (1) and double (2) support phases of walking.

The activity of the various groups of lower limb muscles during walking is shown in **9.5**. From this you should identify the principal muscles causing the following active movements of the striding gait.

Movements of the *hip joint* during walking are largely those of flexion and extension. However, in the double support phase when the trailing leg is about to make its thrust the hip is extending; while that of the front leg about to take on a single support role, is flexed (**9.2b**). There is associated torsion at the lower end of the spine so that the pelvis rotates towards the supporting leg. Since the axes of the feet are considered to be parallel with each other in the direction of motion, the forward hip rotates laterally while the thrusting leg rotates medially (**9.3**). The abduction of the pelvis on the femur of the supporting leg is difficult to perceive, but if you palpate gluteus medius just beneath the iliac crest, you will feel it contract during each supporting phase. If gluteus medius and minimus are paralysed, the pelvis drops on the unsupported side and a very abnormal gait ensues (p. 108) in which excessive lateral flexion of the spine compensates in order to keep the head upright.

During the support phase the *knee joint* becomes progressively more extended (**9.2, 9.4**). After the propulsive thrust the knee of the swinging leg is increasingly flexed until the middle of the swing phase to help the swinging foot clear the ground. It then extends again so that it forms a firm prop when the heel touches the ground at heel strike.

At the *ankle joint* plantar flexion forms part of the propulsive thrust; it ends early in the swing phase when the ankle becomes dorsiflexed (extended) to help the foot to clear the ground (**9.2**). Some plantar flexion starts just before or at heel strike to bring the forefoot into contact with the ground, but the ankle passively dorsiflexes through the single support phase as the body swings forward over the fixed foot (**9.2a, b**). During the double support phase active plantar flexion again occurs to generate thrust.

At the *metatarsophalangeal joints* there is passive dorsiflexion as the ankle is plantar-flexed, and then active plantar flexion, particularly of the great toe, as the thrust-off mechanism is completed.

At the *subtalar* and *transverse tarsal* joints, movement is minimal during the support phase. However, when the backward thrust commences the heads of the metatarsals are fixed on the ground and the posterior part of the foot can move on the transverse tarsal joint as the heel is raised from the ground.

Because of the greater breadth of the female pelvis in relation to the length of the stride, women rotate their hips about a vertical axis more than do men when they walk and in some cultures this is accentuated as a sexual attractant. In fact, each person has his or her own characteristic gait which can, moreover, strongly reflect their mood. The differences between individual gaits can be very difficult to analyse in anatomical terms but, such is the importance to us of being able to recognize individuals, we can often recognize a person in the distance by their gait long before we can distinguish their face.

CHAPTER 10

Answers to questions

Upper limb

Seminar 1

1A The principal form of the bone is determined genetically but the extent of ligament and muscle markings is determined by the tensile forces exerted by these tissues on the bone.

1B A clavicle can be 'sided' by recognizing the superior and inferior surfaces, the larger medial end, and the convexity of the anterior border in its medial third (check your own right and left clavicles by feeling the superior and anterior borders).

1C The scaphoid, because it has the greatest contact with the radius.

1D 6.1.4 is abnormal. A large radio-opaque excrescence (bone) is seen on the profile of the medial border of the humerus which is consequently not well defined. The outer cortex is absent in the area of irregularity. This is caused by a tumour of bone-producing cells.

1E The medulla or cavity of the bone is wider and the cortex is thinner in the abnormal part of the shaft. This is due to a cyst within the humerus.

1F The first metacarpal has a secondary centre of ossification at its proximal end, as have the phalanges.

Seminar 2

2A The clavicle usually lies uppermost because the most frequent cause of dislocation of the acromio-clavicular joint is a direct blow to the acromion. Also, the coraco-clavicular ligament would be ruptured.

2B A fall on the point of the shoulder, e.g. in a riding accident.

2C The weight of the upper limb and contraction of deltoid fibres would depress the outer end of the clavicle. The relation of the clavicle to the coracoid is maintained by the strong coraco-clavicular ligament. The sternomastoid muscle would tend to elevate the inner fragment through its attachment to its medial end.

2D The rhomboids pull the scapulae towards the mid-line, thus bracing the shoulders.

2E Latissimus dorsi adducts, extends, and medially rotates the humerus.

2F Pectoralis major adducts and medially rotates the humerus at the shoulder joint and, through the action of its clavicular head, helps to flex the joint.

2G Pectoralis minor pulls the scapula (and therefore the upper limb) forwards (protraction) around the chest wall.

2H By tilting the chin laterally and upward towards the opposite side against resistance, thereby approximating the mastoid process to the sternum. If the muscles on both sides contract the head is pulled forward (protracted) and downward.

Seminar 3

3A Yes. Probably many other joints, especially the sternoclavicular. Any movement of the shoulder will require postural adaptation of other parts of the body to maintain balance.

3B No, the arm cannot be raised vertically above the head if the scapula is fixed.

3C Synovial fluid lubricates the joint, distributes pressure within it, and helps to nourish the hyaline articular cartilage. When pressure on synovial fluid is increased cross-linkages within its glycoprotein molecules decrease and the fluid becomes less viscous and vice versa (the principle of 'non-drip' paint).

3D Shoulder dislocation initially occurs antero-inferiorly into the dependent, least well-supported part of the capsule. Thereafter the head comes to lie either inferior to the coracoid process anteriorly or beneath the spine of the scapula, posteriorly. Very occasionally it might dislocate upwards with the arm in a vertical position.

3E The axillary nerve, since it lies just beneath the dependent part of the capsule.

3F Subscapularis adducts and medially rotates the humerus. In particular it helps stabilize the shoulder joint as part of the 'rotator cuff'.

3G Teres major and the tendon of latissimus dorsi in addition to subscapularis form the posterior axillary wall.

3H Supraspinatus initiates abduction of the arm at the shoulder joint.

3I Deltoid is the 'power' abductor but cannot initiate the movement; serratus anterior and trapezius rotate the scapula around the chest wall to increase the effective movement of the humerus; the rotator cuff muscles stabilize the shoulder joint; as does the long head of biceps.

3J Extension of the humerus.

3K Deltoid abducts the arm.

3L The humerus has become dislocated from the glenoid fossa and the 'hard lump' that can be felt is the head of the humerus.

3M The subacromial bursa and the tendon of supraspinatus.

Seminar 4

4A The elbow joint has been dislocated. The lower end of the humerus has been displaced forward from the trochlear notch of the ulna, and the radio-humeral joint is also dislocated.

4B Brachialis flexes the elbow joint. Coracobrachialis flexes and adducts the arm at the shoulder joint.

4C Biceps is a strong flexor of the elbow joint with the forearm in the supine position and when acting against resistance, but it contributes much less when the forearm is prone. By its attachment to the radius it is also a powerful supinator with the forearm in the 'position of function'—i.e. elbow semiflexed.

4D Triceps extends the forearm at the elbow joint.

4E Triceps.

4F Biceps and brachialis, the flexors of the forearm, acting against gravity.

4G Normally, if you grasp the medial and lateral epicondyles of the humerus with your thumb and middle finger from posteriorly, your index finger will rest easily on the tip of the olecranon, thus forming the third point of an equilateral triangle. If you cannot do this then the elbow is dislocated. In a supracondylar fracture the triangle would be maintained.

Seminar 5

5A Supination; therefore screws always have right-hand threads. Remember that biceps is a powerful supinator in addition to the supinator muscle.

5B Because the wrist joint abducts or adducts to readjust to the central axis of the forearm and hand.

5C The normal axis passes through the middle finger. To achieve this the lower end of the ulna moves slightly laterally, largely due to the contrac-

tion of anconeus. This enables a grip to be maintained on fixed objects (e.g. a door knob) while the forearm and hand are rotating.

5D At the wrist joint, movements at the radiocarpal and intercarpal joints combine to permit flexion (90°) and extension (45°), abduction (10°) and adduction (45°), and circumduction (a mixture of the other movements) to occur, but no rotation.

5E During wrist adduction the trapezium and the lunate articulate with the radius; during abduction the lunate mostly articulates with the intra-articular disc.

Seminar 6

6A Acting as prime movers: pronator teres pronates the forearm and, by its humeral attachment, contributes to elbow flexion. Flexor carpi radialis flexes and abducts the hand at the wrist joint.
Palmaris longus tenses the palmar aponeurosis and helps to flex the wrist.
Flexor carpi ulnaris adducts and flexes the hand at the wrist joint.
Flexor digitorum superficialis flexes the wrist, the metacarpo-phalangeal joints, and the proximal interphalangeal joints.
All of these muscles contract appropriately in circumduction; and in postural fixation of the wrist to provide a stable support for finger movement.

6B To test flexor digitorum superficialis lay the hand flat on its dorsum and prevent the index, ring, and little fingers from moving; now try to flex the middle finger. It will flex at the metacarpo-phalangeal joint and the proximal interphalangeal joint, but not at the distal interphalangeal joint.
To test flexor digitorum profundus lay the hand in the same position, restrain the middle finger by pressure on its middle phalanx, and again try to flex it. Flexor digitorum profundus will flex the distal interphalangeal joint.

6C Yes, because biceps also acts to supinate and pronator teres to flex.

6D Virtually all: the glass is gripped, the wrist stabilized, and its position delicately adjusted as the elbow and shoulder joints are flexed against gravity.

6E Because if it were, pronation and supination would be limited.

Seminar 7

7A The position of the fingers and thumb enables objects to be grasped readily; and enables the thumb to oppose each of the fingers for purposes of grip and manipulation.

7B Flexion (90°), extension (5°), abduction and

adduction (40°) can occur at the metacarpo-phalangeal points, while a combination of these movements produces circumduction. Only flexion (90°) and extension (which returns the flexed joint to the anatomical position, but not beyond) occurs at the interphalangeal joints.

7C It helps to provide Man with the ability to manipulate even very small objects with exquisite precision.

7D It might have involved parallel evolution of mechanisms concerned with the conduction and modification of sensory information reaching the brain; conscious appreciation of what is being held and the use to which it can be put; correlation, integration, and modification of appropriate behavioural and motor responses.

7E The lumbrical tendons cross anterior to the metacarpo-phalangeal joints (and will therefore flex them) to insert into the dorsal extensor expansion of the fingers. Therefore the lumbricals also extend the interphalangeal joints.

7F If the synovial tendon sheath of the little fingers is infected, the infection might well spread to the common flexor synovial sheath with which it is connected. This is a serious development since fibrous adhesions can form between the inflamed tendon sheath and the tendons, thus limiting their movement.

7G An incision into the web between the thumb and index finger (**6.7.16**, arrow) permits drainage of the thenar space. An incision into the web between the middle and ring fingers (**6.7.16**, arrow) permits drainage of the mid-palmar space.

Seminar 8

8A An outer layer of connective tissue containing vessels and nerves (tunica adventitia); a middle layer of smooth muscle and elastic tissue (tunica media); and an inner connective tissue and endothelial layer (tunica intima).

8B An artery transmits the pulsations of the blood derived from the contraction of the ventricles of the heart (systole).

8C A deep vein or veins is connecting with the superficial vein.

8D First, by direct pressure on the palm of the hand (after first checking that no sharp object is still embedded there) and by elevating the limb. If this fails, then compression of the artery proximal to the injury is required. A tourniquet can be applied to the upper arm, or the axillary or subclavian arteries can be compressed. It is essential that a tourniquet be applied for a minimum period only (about 30 min.), or tissue death might occur.

8E To maintain the higher blood pressure in the arteries and smooth the pulse pressure.

8F Lymph from the little finger drains along the course of the basilar vein to nodes in the axilla; possibly a gland close to the medial epicondyle would also be enlarged.

8G The arterial obstruction lies at the junction of the right subclavian artery with the common carotid artery.

Seminar 9

9A Serratus anterior would be incapable of holding the scapula close to the chest wall and of pulling it around the chest wall; forward thrusting movements would be weak and the scapula would 'wing' if they were attempted.

9B Abduction of the humerus at the shoulder joint would be weak (supraspinatus); extension of the shoulder likewise (infraspinatus).

9C To redistribute the motor and sensory fibres of the nerve roots.

9D Sensory loss (C7 dermatome) over the front and back of the index, middle, and ring fingers, and the middle part of the front and back of the hand and wrist.

9E The scapula would be 'winged' (serratus anterior); adduction would be weak (pectoralis major, latissimus dorsi); the limb would be mottled in colour because of lack of control of blood flow to the skin, and sweating reduced (autonomic nervous system).

Seminar 10

10A Muscles of the anterior compartment of the arm (biceps, brachialis, coracobrachialis) would be paralysed and flexion at the shoulder and especially the elbow would be weak. There would be sensory loss over the lateral aspect of the forearm.

10B Anaesthetic should be injected into the skin on the medial and lateral aspects of the root of the fingers (ring block) to reach the digital nerves. Catecholamines can be added to local anaesthetics to reduce blood flow and thereby localize their action; the smooth muscle of the arteries may be hypersensitive to the amines and this could lead to arterial spasm and cutting off the blood supply to the fingers.

10C Paralysis of the muscles of the anterior compartment of the forearm with the exception of flexor carpi ulnaris: therefore flexion at the wrist, metacarpo-phalangeal, and interphalangeal joints, and abduction at the wrist will be weak.

Seminar 11

11A Because the lateral two lumbricals (which flex the metacarpo-phalangeal joints and extend the interphalangeal joints of the middle and index fingers) are supplied by the median nerve.

11B   The ulnar nerve in the forearm supplies flexor carpi ulnaris and that part of flexor digitorum profundus which inserts into the little and ring fingers. Therefore damage to the nerve at the elbow would lead to weak ulnar deviation at the wrist and weak flexion of the little and ring fingers in addition to the defects found after severance of the nerves at the wrist (see text).

Seminar 12

12A   The muscles of the anterior wall of the axilla (pectoralis major and minor) are supplied by medial and lateral pectoral nerves of the medial and lateral cords of the brachial plexus. Serratus anterior, which forms the medial wall, is supplied by the long thoracic nerve from the roots of C5,6,7.

12B   Wrist drop and weakness of finger extension would be the most marked features owing to paralysis of the muscles of the posterior compartment of the forearm. Triceps would probably not be impaired because its nerves arise in the axilla and upper arm. Sensory loss in the forearm might be difficult to detect because the supply of the medial and lateral cutaneous nerves of forearm overlap on to its posterior surface, but some loss over the dorsum of the radial side of the hand should be detectable.

Lower limb

Seminar 1

1A   The neck of the femur separates the femoral head from its shaft and greater tuberosity, thus reducing limitation of movement at the hip joint. The role of the clavicle is comparable in that it acts as a strut by which the upper limb is suspended clear of the trunk to enable a wide range of movements to take place at the shoulder.

1B   The wide platform-like surfaces of the tibia, which articulate with the condyles of the femur, provide for transmission of the weight of the limb to the lower leg in addition to movements at the knee joint.

1C   The arch of the foot, formed by several small bones (well supported by ligaments and muscles) transmits the weight of the body to the ground and provides flexibility while moving over uneven ground.

1D   The collagenous fibres on which crystals of bone salts (calcium hydroxyapatite) are deposited are laid down along lines of stresses and strains within the bone. Many of these forces result from weight-bearing, though the basic trabecular architecture is genetically determined. Note, in particular, the spur of bony trabeculae (calcar femorale)

that extends from the upper aspect of the femoral head obliquely down to the inner aspect of the femoral shaft.

1E   The hip joint is normal on the right side. The femoral head on the left appears to be crumbling and a line drawn along the upper margin of the neck of the femur does *not* pass through the upper third of the capital epiphysis.

Seminar 2

2A   The articular facets of the shoulder joint enable a wide range of movement to take place; stability in this joint depends on ligaments and, particularly, on the rotator cuff muscles. The articular facets of the hip joint, on the other hand, provide a much greater degree of stability, commensurate with weight-bearing locomotion, and therefore a smaller degree of movement.

2B   The head of the femur is dislocated on the left side; Shenton's line is discontinuous.

2C   Treatment of congenital dislocation of the hip is to place the infant in a plastic cast with both hips abducted. In Nigeria, babies are often carried astride their mother's hip (i.e. with both legs abducted); therefore congenital dislocation is less often apparent in that country.

2D   It is likely that one or other of the cruciate ligaments have been ruptured. The 'anterior drawer' or 'sag' tests would indicate which of the two has been damaged.

2E   It is most likely that the medial meniscus of the supporting knee will have been torn. The meniscus is held by the medial collateral ligament and is subject to rotational stress as the momentum of the body attempts to rotate the femur on the supporting tibia.

2F   If the injury results from hyper-inversion of the foot then lateral structures of the ankle may be damaged—the lateral ligament (3 components) and the lateral malleolus; if the injury results from hyper-eversion then medial structures of the ankle may be damaged—i.e. the deltoid ligament and the medial malleolus. (N.B. There are other possible causes.)

Seminar 3

3A   It is active in strong lateral rotation of the thigh and in tension of the iliotibial tract and thus may have an action on the knee joint, especially when standing on one leg.

3B   The combined action of these three muscles is to stabilize the hip joint; they can also laterally rotate the hip joint.

3C   Iliacus and psoas are powerful flexors of the thigh on the trunk (or the trunk on the thigh).

3D If the neck of the femur is fractured the shaft of the femur is freed from the restraint of the hip joint; psoas pulls forward on the lesser trochanter of the mobile shaft and therefore laterally rotates the shaft (see **7.2.2**).

Seminar 4

4A Adductor longus contracts on medial rotation of the hip joint.

4B All the hamstrings extend the hip joint; with the exception of adductor magnus, they also flex the knee.

4C All muscles of the quadriceps group extend the knee; rectus femoris also stabilizes the hip joint and assists ilio-psoas in flexing it.

4D The attachment of vastus medialis to the medial aspect of the patella helps to prevent a tendency to lateral displacement of the patella, which results from the angulation of the femoral shaft on the tibia.

4E Rectus femoris has ruptured.

4F The adductor muscles can also medially rotate the thigh; adductor magnus acting as a hamstring extends the thigh; gracilis flexes and can medially rotate the knee joint.

4G The pubic tubercle can be easily located by palpating the tendon of adductor longus and following it upward to its attachment to the pubis immediately below and medial to the pubic tubercle.

4H Adductor magnus has both adductor and hamstring components. Therefore it is supplied by both the obturator and sciatic nerves.

4I Popliteus acts to 'unlock' the fully extended knee joint by laterally rotating the femur on the tibia.

Seminar 5

5A The anterior compartment muscles dorsiflex (extend) the foot at the ankle, while tibialis anterior also inverts the foot and helps to support the medial aspect of the transverse arch.

5B Peroneus longus and brevis evert the foot and plantar-flex (flex) it.

5C The superficial calf muscles plantar-flex the foot, thereby enabling a person to stand on the ball of the foot (or the tips of the toes). Soleus is more of a 'postural' muscle, while gastrocnemius wins the long jump!

5D *a*) No effect except that plantar flexion would be weaker.
b) Inability to propel the body forwards as in walking.

Seminar 6

6A 'Full' medial and lateral rotation of the lower limb takes place only at the synovial ball and socket hip joint.

6B With the full weight of the body being transmitted forward to the toes and tension of the small muscles of the sole being reduced, the medial longitudinal arch of the foot must increase.

6C Flexion and extension take place at the ankle joint (a synovial hinge joint); inversion and eversion take place at the subtalar (and transverse tarsal) joints.

6D Flexion, extension, and slight abduction and adduction take place at the metatarso-phalangeal joints (synovial condyloid joints); only flexion and extension occur at the interphalangeal joints (synovial hinge joints).

6E The fibro-fatty tissue of the heel provides a resilient pad which absorbs forces transmitted between the heel and the ground—i.e. it is an excellent shock-absorber.

6F Tendons of flexor digitorum superficialis split to allow the tendons of flexor digitorum profundus to pass through to their insertions at the base of the distal phalanges.

6G The gripping action of the toes helps to stabilize the balance of the body when standing upright; and facilitate the forward push-off during locomotion.

6H The medial plantar nerve is comparable in its distribution to the median nerve.

Seminar 7

7A With the attainment of the upright stance venous return to the heart from the superficial veins of the lower limb has to travel a greater vertical distance against gravity. The valves and the communicating veins which allow blood to drain from the superficial to the deep veins (which are emptied by the muscle pump) aid venous return.

7B The deep palmar and deep plantar arches are comparable but, in the foot, there is no equivalent of the superficial arch, presumably because of the pressure exerted by the body weight during standing.

7C Insufficient blood supply. The extent of any pulsations in the main arteries to the limb (e.g. the tibial artery behind the medial malleolus) should be determined and the cause of the problem investigated. Pharmacological inhibition or surgical interruption of the sympathetic nerve supply may help; vascular bypass grafts are sometimes possible. However, amputation of the devascularized area may become necessary.

7D No, since there is no tightly confined osteo-fascial compartment in the calf.

Seminar 8

8A   S3, 4; note that this reflects the caudal origin of this area . The lower limbs develop cranial to this region, as lateral extensions of the trunk, and subsequently rotate to their definitive position.

8B   L2, 3, 4, sensory nerve roots are stimulated. Sensory (afferent) fibres synapse with motor (efferent) neurons at spinal cord levels L2, 3, 4, thus forming a monosynaptic reflex arc. The knee jerk is elicited in order to test the integrity of the sensory and motor fibres which comprise the arc and especially their excitability, which depends on influences from higher nervous centres. The reflex response on one side must always be compared with that on the opposite side.

8C   The obturator nerve in the pelvis lies just lateral to the ovary. Rupture of follicles at ovulation may irritate the adjacent nerve.

Seminar 9

9A   Foot drop with wasting and loss of extension (dorsiflexion) and eversion due to paralysis of anterior and lateral compartment muscles; loss of sensation over the lower lateral part of the front of the leg and the dorsum of the foot.

9B   Paralysis and wasting of hamstring and calf plantar muscles (tibial nerve), in addition to the muscles of the anterior and lateral compartments of the leg (common peroneal nerve); almost complete loss of sensation below the knee. A major disability would be foot drop, which can be helped very simply by attaching a spring to the front of the shoe which is secured to a band below the knee.

Spine

1A   Growth of the vertebral column during development elongates it more (in a rostrocaudal direction) than does that of the spinal cord.

1B   The side on which the leg is lifted. Erector spinae contracts to laterally flex the spine away from the supporting leg and thus restore the centre of gravity which has been disturbed.

1C   At the atlanto-occipital joint, nodding movements are possible with slight lateral flexion but no rotation; at the atlanto-axial joint, the atlas rotates around the dens of the axis.

The intercervical articular facets lie on a sloping, transverse plane which permits flexion/extension and a small amount of lateral flexion.

The interthoracic articular facets lie on the circumference of a large circle, the centre of which is in the trunk, thus permitting rotation, but rather little flexion or extension, which is limited by the ribs.

The interlumbar facets also lie on the circumference of a circle but, because the centre of that circle is dorsal to the trunk, the facets form strong interlocking articulations which permit some flexion and extension but no rotation of the trunk.

1D   Accessory cervical ribs give rise to more problems. T1, which forms the lower part of the brachial plexus, has to emerge from the chest and travel over the first rib to reach the plexus. If an accessory rib is present it passes over this and is stretched, especially when the arm is carrying a heavy weight.

1E   The transverse ligament of the atlas prevents the odontoid peg from moving backwards and therefore damaging the medulla and upper cervical spinal cord.

1F   The alar ligaments restrain both rotation and flexion/extension between the occiput and the axis.

1G   To open up the area between the 4th and 5th lumbar spines where the lumbar puncture needle is being inserted.

1H   Paralysis and loss of sensation below the level of the lesion. Autonomic responses would lack higher control—therefore loss of coordination of pelvic visceral reflexes.

The patient would have the use of the diaphragm (C3, 4, 5) and could therefore breathe.

C5, 6 remain intact: therefore most shoulder joint movements would be possible, though weaker, abduction of the shoulder being most affected. At the elbow, flexion would be possible and weak flexion of the wrist might also be possible.

1I   Diaphragmatic and intercostal breathing would be intact, as would sympathetic reflexes.

1J   This lesion, which is relatively common, could give rise to pain down the lower and lateral aspect of the leg, passing into the foot (S5 dermatome). It would also lead to weakness of eversion of the foot.

Appendix

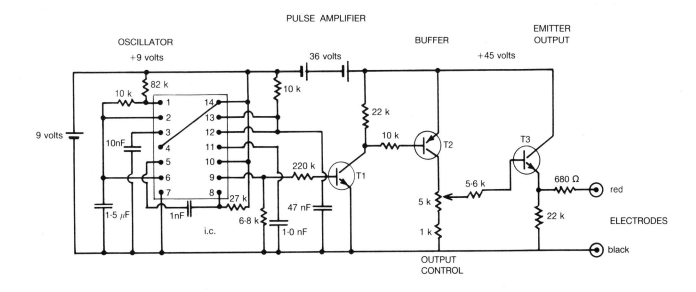

Fig. A.1   Circuit diagram of an electronic stimulator for testing muscles. (T1: ZTX304; T2: ZTX504; T3:ZTX304; i.c.: NE556A.)

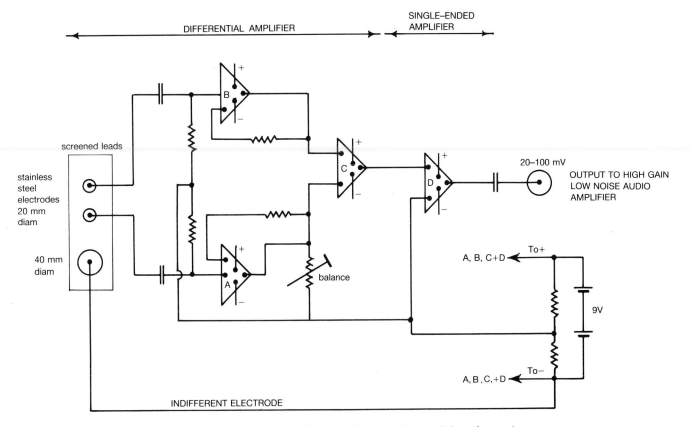

Fig. A.2   Block diagram of a simple electronic amplifier for the electrical activity of muscles.

Index

Illustrations are indicated by italic page numbers after a semicolon.